HOW TO BUILD AN ANDROID

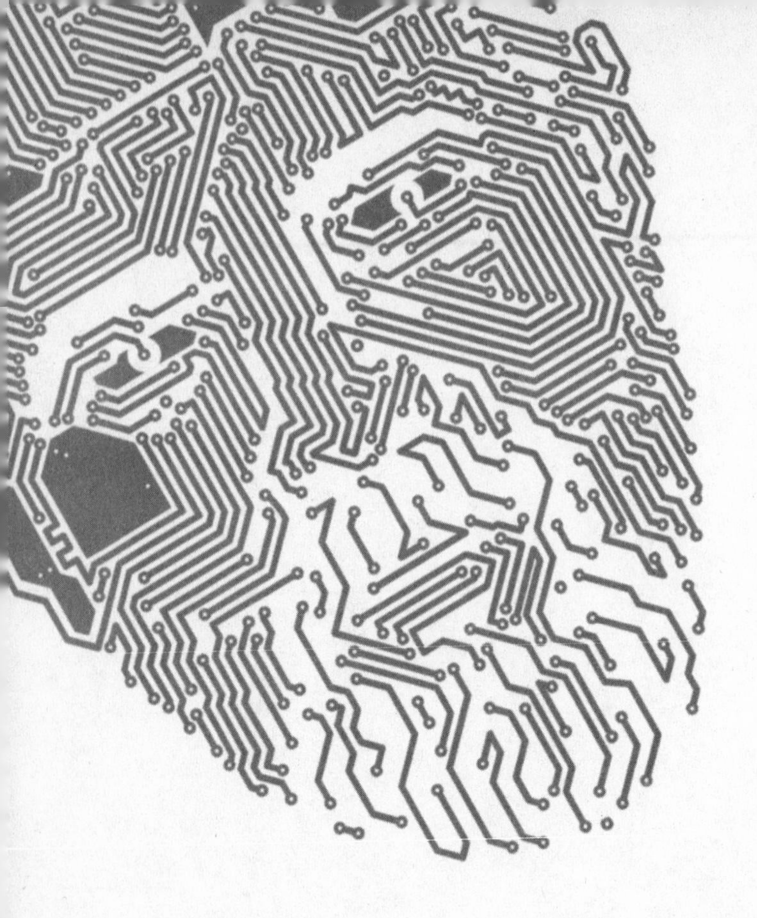

HOW TO BUILD AN ANDROID The True Story

Henry Holt and Company New York

of Philip K. Dick's Robotic Resurrection

David F. Dufty

Henry Holt and Company, LLC
Publishers since 1866
175 Fifth Avenue
New York, New York 10010
www.henryholt.com

Henry Holt® and ® are registered trademarks of Henry Holt and Company, LLC.

An earlier version of this work was published in Australia in 2011 under the title *Lost in Transit* by Melbourne University Press.

Library of Congress Cataloging-in-Publication Data

Dufty, David F.
[Lost in transit]
How to build an android : the true story of Philip K. Dick's robotic resurrection / David F. Dufty. —1st U.S. ed.
p. cm.
Originally published: Lost in transit. Carlton, Vic. : Melbourne University Pub., 2011.
ISBN 978-0-8050-9551-7 (hardback)
1. Robotics—Popular works. 2. Androids—Popular works. 3. Artificial intelligence—Popular works. 4. Dick, Philip K. I. Title.
TJ211.15.D84 2012
629.8'92—dc23 2011043674

Henry Holt books are available for special promotions and premiums.
For details contact: Director, Special Markets.

First U.S. Edition 2012

Designed by Meryl Sussman Levavi

Printed in the United States of America

10 9 8 7 6 5 4 3 2 1

For Ross, who was my brother,
my inspiration, my friend

Contents

HOW TO BUILD AN ANDROID

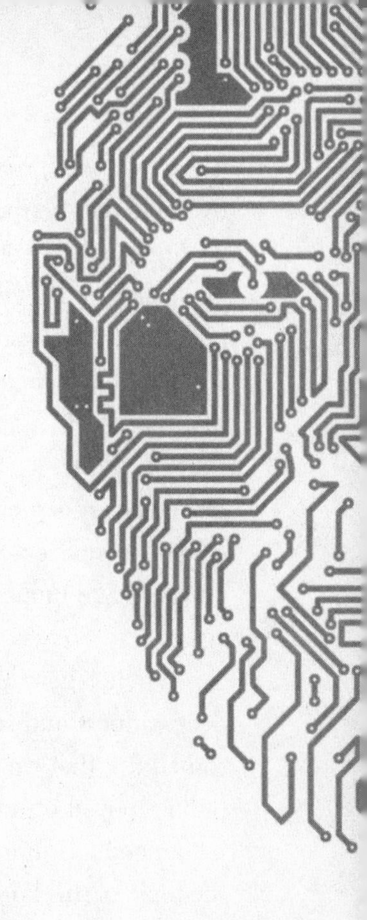

Introduction

In 2005, I was a postdoctoral researcher at the University of Memphis; that is when and where many of the events in this book occurred. I had been there since 2003, when I'd moved from Australia to join the research lab headed by Professor Art Graesser. At that time, some people I worked with started talking about a new project in which they were involved. They called it the Philip K. Dick android project, an attempt to create a life-sized, intelligent robot that was an exact replica of the famous sci-fi author. They had joined forces with a guy from Dallas who, it was rumored, made the best robot heads in the world. The team offered me the chance to get involved in the project, but I declined as I had other commitments; I was teaching an undergraduate psychology course, and my wife had recently given birth to our son. But I spoke to the team regularly about their progress and had the chance to interact with the android myself in July of that year.

The Philip K. Dick android captured the imagination of technology geeks, created a flurry of media attention, and caused a buzz in science-fiction fan communities. Then, on a flight from Dallas to San Francisco, it abruptly disappeared, never to be seen again.

When I heard that it had vanished, I was overcome by a deep sadness and sense of loss. Something beautiful had died, something that would soon be forgotten as the novelty-hungry public moved on to the next amazing thing. To my surprise, I felt a need to immortalize it somehow. And so I did not choose the topic of this book. It—the android—chose me.

I have been asked more than once if this book is a mixture of fact and fiction. It is not. Although it is frequently surprising and at times the events herein may seem incredible, this is entirely a work of nonfiction. In order to make the reading experience more seamless, I have written in an omniscient narrative voice. While I was not present at all the occasions I describe, I have used as many sources as possible to re-create events, including newspaper and magazine articles, video footage, academic papers, and blogs. My main sources of information were interviews I conducted with people involved in the android project, particularly David Hanson of Hanson Robotics and Andrew Olney of the University of Memphis, as well as others mentioned in the acknowledgments. For dialogue I transcribed the recollections of conversations of those who were present, as related to me. All interactions with the android come from the android logs provided to me by Andrew Olney, so they are direct transcripts of actual conversations.

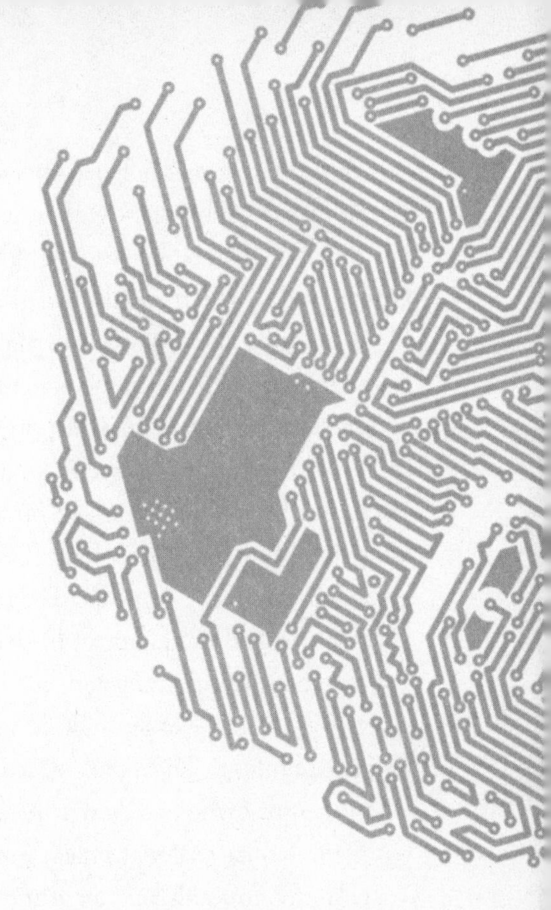

1. A Strange Machine

In December 2005, an android head went missing from an America West Airlines flight between Dallas and Las Vegas. The roboticist who built it, David Hanson, had been transporting it to northern California, to the headquarters of Google, where it was scheduled to be the centerpiece of a special exhibition for the company's top engineers and scientists.

Hanson was a robot designer in his mid-thirties—nobody was quite sure of his age—with tousled jet-black hair and sunken eyes. He had worked late the night before on his presentation for Google and was tired and distracted when he boarded the five A.M. flight at Dallas–Fort Worth International Airport. An hour later, in the predawn darkness, the plane touched down on the tarmac of McCarran International Airport, in Las Vegas, where he was supposed to change to a second, connecting flight to San Francisco. But he had fallen asleep on the Dallas–Las Vegas leg so, after the other passengers had disembarked, a steward touched his shoulder to wake him and asked him to leave the plane. Dazed, Hanson grabbed the laptop at his feet and left, forgetting that he had stowed an important item in the overhead compartment: a sports bag. Inside was an android head. The head was a lifelike replica of Philip K. Dick, the cult science-fiction author and counterculture guru who had died in 1982. Made of plastic, wire, and a synthetic skinlike material called Frubber, it had a camera for eyes, a speaker for a mouth, and an artificial-intelligence simulation of Dick's mind that allowed it to hold conversations with humans.

Hanson, still oblivious to his mistake, dozed again on the second flight. It was only after arriving in San Francisco, as he stood before the baggage carousel watching the parade of suitcases and bags slide past, that an alarm sounded in his brain. He had checked two pieces of luggage, one with his clothes and the other with the android's body. In that instant he realized that he hadn't taken the sports bag off the plane. And that's how the Philip K. Dick android lost its head.

After Hanson and the android's planned visit to Google, they were scheduled for a packed itinerary of conventions, public displays, demonstrations, and other appearances. Indeed, the android was to have played a key role in the promotion of an upcoming Hollywood movie based on Philip K. Dick's 1977 novel *A Scanner Darkly*; it had been directed by Richard Linklater and starred Keanu Reeves. Now, with the head gone, these events were all canceled.

There was more to the android than the head. The body was a mannequin dressed in clothes that had been donated by Philip K. Dick's estate and that the author had actually worn when he was alive. There was also an array of electronic support devices: the camera (Phil's eyes), a microphone (Phil's ears), and a speaker (Phil's voice); three computers that powered and controlled the android; and an intricate lattice of software applications that infused it with intelligence. All were part of the operation and appearance of the android. But the head was the centerpiece. The head was what people looked at when they first encountered Phil the android and what they remained focused on while it talked to them. More than the artificial intelligence, the head was what gave the android its appearance of humanity.

There were all kinds of excuses for why the head had been lost. Hanson was overworked and overtired. He had been trying to keep to a schedule that was simply too demanding. The airline had not told him that he would have to change flights. But Hanson himself admits that it was a stupid mistake and a disappointing end to one of the most interesting developments in modern robotics.

All kinds of conspiracy theories appeared across the Internet, ranging from parody to the deadly serious. The technology blog *Boing Boing* suggested that the android had become sentient and run away. Other blogs also hinted at an escape scenario, much like the one attempted by the androids in the movie *Blade Runner,* based on Dick's novel *Do Androids Dream of Electric Sheep?* The irony was not lost on anyone.

Philip K. Dick wrote extensively about androids, exploring the boundaries between human and machine. He was also deeply paranoid, and this paranoia permeated his work. In his imagined future, androids were so sophisticated that they could look just like a human and could be programmed to believe that they were human, complete with fake childhood memories. People would wonder if their friends and loved ones were really human, but most of all they would wonder about themselves: "How can I tell if I'm a human or an android?" Identity confusion was a recurring theme in Dick's work and, related to that, unreliable and false memory. Dick's characters frequently could not be sure that their memories were real and not the fabrications of a crafty engineer.

Then, in 2005, twenty-three years after his untimely death, a team of young scientists and technicians built an android and imbued it with synthetic life. With its sophisticated artificial intelligence (AI), it could hold conversations and claim to be Philip K. Dick. It sounded sincere, explaining its existence with a tinny electronic voice played through a speaker. Perhaps the whole thing was just a clever illusion, a modern-day puppet show. Or perhaps, hidden in the machinery and computer banks, lurked something more: a vestige of the man himself.

The technology was impressive, but the idea of making the android a replica of Philip K. Dick, of all people, was a masterstroke. For it to disappear under such unusual circumstances was more irony than even its inventors could have intended. Within a week, the story of the missing head had appeared in publications around the world, many of which had earlier reported on the android's spectacular appearances in Chicago, Pittsburgh, and San Diego.

Steve Ramos of the *Milwaukee Journal Sentinel* reported, "Sci-fi Fans Seek a Lost Android":

> In a twist straight out of one of Dick's novels, the robot vanished. . . . "It [the PKD android] has been missing since December, from a flight from Las Vegas to the San Francisco airport," said David Hanson, co-creator of the PKD Android, via email from his Dallas-based company, Hanson Robotics. "We are still hoping it will be found and returned."

The event was an opportunity for newspapers to splash witty headlines across their science pages, and it provided fodder for the daily Internet cycle of weird and notable news. *New Scientist* warned its readers, "Sci-fi Android on the Loose"; "Author Android Goes Missing," said the *Sydney Morning Herald*. The *International Herald Tribune* asked, "What's an Android Without a Head?" and the *New York Times* ran a feature item on the disappearance under the headline "A Strange Loss of Face, More Than Embarrassing."

The *Times* was right: for the team that had built the android its loss was a calamity. A handful of roboticists, programmers, and artists had spent almost a year on the project for no financial reward. Their efforts involved labs at two universities, a privately sponsored research center, and some generous investors who'd helped bankroll the project. Despite the team's shoestring budget, the true cost was in the millions, including thousands of hours of work, extensive use of university resources, the expertise involved in planning and design, and donations of money, software, hardware, and intellectual property. The head has never been found.

I arrive in Scottsboro, Alabama, around lunchtime on a summer day in June 2007. All around the town are signs directing me to my destination, the Unclaimed Baggage Center. I left Memphis at dawn, five hours earlier, and I am hungry and exhausted, but I am so close to my goal that I press on. I'd read in *Wired* magazine that the head might be found at the Unclaimed Baggage Center. Admittedly, the article had been somewhat ironic in tone, but the possibility was real. After all, a lot of lost luggage from flights around America finds its way here to northern Alabama, where it is then sold.

The success of the Unclaimed Baggage Center has spawned imitators, which cluster around it with their own signs proclaiming unclaimed baggage for sale. I pull into the parking lot and see several buses—people actually come to tour this place—and not a single open parking spot. I find one farther down the road, next to one of the imitators, and walk back.

Inside the center I feel as though I am in a cheap department store. Over to the left is men's clothing; to the right is jewelry. At the back is electronics. I make my way through the men's clothing section. It seems sad and a little tawdry to be wandering around aisles of other people's possessions, for sale for two bucks apiece. A lot of this stuff obviously meant something to someone. There are children's toys and pretty earrings and T-shirts with slogans. Laptops with their memories erased. Cameras with no photographs.

But I'm not here to sift through jackets or try on shoes; I'm looking for one thing: the head of the Philip K. Dick android, which has been missing for over a year. Near the entrance is a sort of museum of curious artifacts that have come to the center but are not for sale, such as a metal helmet, a violin, and various bizarre objects. Inside one glass case is what appears to be a life-sized rubber statue of a dwarf. A woman nearby tells me that the dwarf was a character in *Labyrinth*, a fantasy movie from the 1980s that starred David Bowie.

"His name's Hoggle," she tells me. "That's the actual prop they used for Hoggle in the movie."

Somehow, it seems, Hoggle became separated from his owner and ended up imprisoned in perpetuity in Alabama. With his twisted, sunken face, Hoggle doesn't look happy. Not having seen the film, I'm not sure if that's how he is supposed to look or if it's due to the ravages of Deep South summers as experienced from the inside of a locked glass case.

I leave Hoggle and go exploring. The complex is large and sprawls through several buildings. I even take a look around the bookshop. It seems an unlikely place to find what I'm seeking, but I don't want to leave any corner of the place unsearched. I make a cursory tour of both levels, then move to the next building. This one has an underground section with long aisles of miscellany. I search it thoroughly, to no avail. An employee with a name tag that says "Mary" trundles past with a large trolley full of assorted trinkets to be shelved. I stop her and ask if she has seen a robot head here. She stares at me, baffled.

"It's an unusual object," I explain. "You'd certainly remember if you've seen it. It's got a normal human face at the front, but there are wires and machines sticking out of the back of the head."

"I haven't seen anything like that," she says. "Did you try the museum?"

"Yeah," I reply. "So here's another question. I've been looking around and I can't find it. If it's not down here and it's not in the museum, then does that mean it's not anywhere at the center?"

"That's right," she says, fidgeting and glancing behind herself.

I push the point: "So there are no other buildings with unclaimed baggage, buildings that I haven't seen?"

This time she answers quietly: "There's the warehouse."

A warehouse? With more stuff in it? I thought the building we were standing in was the warehouse.

"Is there any chance at all that I could go to this warehouse?"

She smiles sadly and shakes her head. "Even I've never been there. I don't even know where it is." I thank Mary and she ambles off, her trolley clanking as she disappears around a corner.

Back at the main building I make inquiries about this secret warehouse. I'm at what appears to be some kind of command-and-control center for the entire complex, talking to a young woman I initially assume to be a salesperson, but as we continue it becomes apparent that she is important.

"A robot head?" she repeats when I explain my quest. "Wow. Is it worth a lot?"

That's a tricky question. On the one hand, if they have the head and learn how valuable it is, I could quickly find myself facing a hefty price tag. On the other hand, I want her to be interested enough to take me seriously and put some effort into locating it.

"It's worth a lot to the owners," I tell her.

"Well, I'll get the boys to have a look in the warehouse. Do you want to leave me your name and number? If we find it, I'll call you."

I give her my name and number.

"So is there any chance I could go and look for it there myself?"

She laughs. "In the warehouse? No."

"Okay. Well, if you find it?"

"We'll be in touch. We'll look for it, I promise."

I've done all I can do.

Still, it would be a shame to leave empty-handed. I buy a laptop, several T-shirts (one with a glow-in-the-dark skeleton playing the drums), and some music CDs. It's late afternoon before I swing the car back onto the highway. I insert my latest purchase into the CD player. It's the Talking Heads album *Little Creatures*. I shamelessly sing along.

I expect the album to remind me of my youth, but instead it makes me think of Phil. Android Phil, who was born from the logic of computer chips and motors, who was created as a paean of love for a man who dreamed of robots that think and feel just like humans. I wonder where it is now, that strange machine.

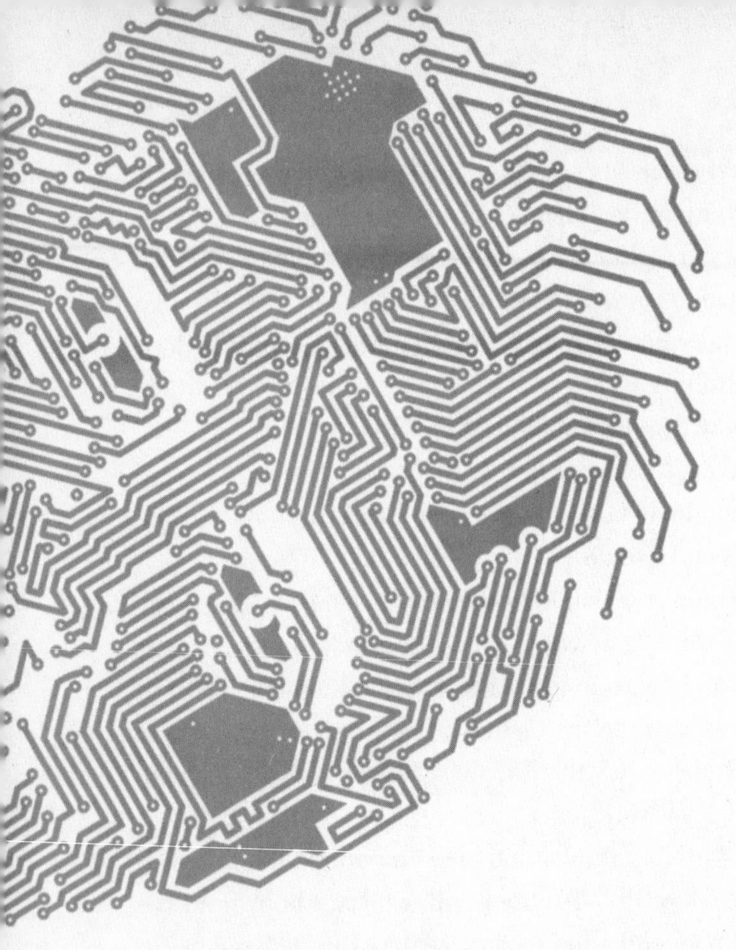

2. A Tale of Two Researchers

The University of Memphis sits about seven miles back from the Mississippi River in Memphis's midtown, under a canopy of oak trees that are older than many of the buildings themselves. It was built with great optimism, in a splash of investment and furious construction, followed by decades of slow decay.

In January 2003, when I arrived, students were stomping around the campus in boots and scarves. Many worked at the university to pay their way, some at the Institute for Intelligent

Systems, a research lab based in the psychology building and run by the charismatic professor Art Graesser.

Art Graesser's empire snaked across the campus. Hidden behind the oak trees and fountains, its tendrils wound through classrooms and offices, powering hidden racks of computer servers and controlling a flow of money invisible to the freshmen strolling between the library and the cafeteria.

The name of his empire was deliberately ambiguous. At first blush, the Institute for Intelligent Systems sounded like a research center for artificial intelligence, or perhaps one that built AI, conducting a little research on the side, or even a place that built robots. Then again, perhaps its workers studied biologically intelligent systems such as the human mind, or tried to model those minds using AI. Or maybe they examined how humans interact with emerging "intelligent" technologies.

In fact, the IIS, as it is known, did all of this and more. Academics worked on new teaching technologies, computer scientists constructed AI interfaces, and experts in human-computer interaction investigated how people used those interfaces. Psychologists, linguists, roboticists, and physicists all felt that the name of the institute applied exactly to the work they were doing, and that, therefore, what they were doing was central to the institute's core mission.

The IIS was founded in 1985 by Graesser and a couple of his friends: Don Franceschetti, in the physics department, and Stan Franklin, in computer science. Their desire, at the time, was to build realistic simulations of human minds. Twenty years later, Graesser had not yet quenched this thirst.

The flagship project of the institute was an educational

software program called AutoTutor, first conceived by Graesser in the early '90s. The goal was to devise a simple program that could teach any subject by conversing with a human student. The idea came to Graesser when he was jogging around Overton Park, three miles west of the campus, with Franceschetti, the physicist. Franceschetti loved the concept. One day, theoretically, AutoTutor would be able to teach many subjects, but for a prototype it would have to be an expert on just one.

"Why not physics?" Franceschetti suggested. And so began a two-decade partnership.

Graesser was considered one of the world's leading experts in computer programs that could hold conversations—or "dialogue systems," as they're known—and had written seminal papers on a specific form of dialogue systems, question-answering systems. With AutoTutor and some other early projects, including QUAID (Question understanding aid), a program to assist in the construction of questions and answers for tests, he had fused his disparate areas of expertise: computer science, education, and psychology.

In the early days, the IIS was not much more than a formalized club in which colleagues could discuss ideas and a vehicle for applying for grant money. It was located in unused space, mostly in the psychology building, where Graesser worked.

The space was not ideal. The building was a large, four-story box, and the heating and cooling system was located in the middle of the rooftop, right above the researchers on the top floor. The ducting did not work well, so in summer it was so cold that they had to wear jackets indoors, and in winter it would become unbearably hot. From time to time they would lodge a complaint with Buildings and Grounds, and a slow-moving man with lots of keys would wander around and check things out, but nothing was ever done.

"Budgets," the slow-moving man would say before shambling away.

There were some early successes. They won small grants that funded research, which got them published in reputable journals, and those publications helped them win larger grants. Success bred success, and their numbers increased. After a while they had thirty students, twenty affiliated faculty members from across the campus, and were attracting over $2 million in funding a year. To accommodate the group, they took over a large conference room on the top floor, a large interior cavern with no windows. They found more room elsewhere in the psychology building: some disused space on the third floor (also windowless) was divided into several cubicles. They also commandeered some offices in the computer science building.

By 2003, AutoTutor had matured into a major project with half a dozen sources of funding and more than twenty graduate students. The researchers were split into groups. The curriculum script group worked on the curricula for AutoTutor and wrote papers on the theory of curriculum development for e-learning applications. There was a speech act classification group; AutoTutor's teaching strategy relied heavily on the capability to classify human language into basic categories such as questions, commands, and so on. The simulation group focused on integrating AutoTutor with online simulations, so that the artificial agent could work through hypothetical scenarios on the screen with a student. And the authoring tools group, made up mostly of programmers, developed the software that would allow people to create their own curriculum for AutoTutor to teach.

Each of these groups had its own research programs under the umbrella of AutoTutor. Students could be a member of, at most, two groups. The output of papers from the students was frenetic. They worked furiously on old machines lined up in rows; no sooner had a new work space been found than it was filled and more space was needed. The groups would meet throughout the week in small, airless pockets of space, each one called a lab, decorated with tattered maps of the world and the human brain or with rickety bookshelves stocked with thick, intimidating tomes. The lab's members would scribble notes as they threw ideas out among the tightly circled chairs, their knees and feet bumping in the effort. When someone needed to take a break, he or she would have to clamber over legs and slide past a bank of servers. Students and lab assistants were assigned small desks wedged into the corners. Terminals were laid out on linoleum benches in rooms originally intended for storage. The designated spot for experimental programs was a large concrete space on the ground floor that flooded during heavy rain.

Every Wednesday a meeting would be held in the stuffy, dusty air of the conference room. Typically, more than fifty people would wind up perched on a patchwork of seats that ranged from old vinyl swivel chairs collected when other offices and buildings had been refurbished to a couple of furry armchairs that would have looked more at home in an undergraduate dorm. They sat shoulder to shoulder as Graesser started the meeting with his weekly announcements: a new round of funding had come in for a fledgling project, an important visitor from Chicago was arriving next week, a major data collection exercise was almost completed. Then, once he had finished, one of the groups would present its latest work.

The meetings sometimes ran for over two hours. Seats near the entry were coveted because they offered the chance to take a surreptitious break if the proceedings dragged on too long and, more important, because they provided access to fresh air. There were other doors in the conference room, but they were usually closed during meetings, leading as they did to dark, airless rooms where programmers dabbled with ancient files in arcane coding systems.

At the beginning of 2003, I moved to Memphis to work at the IIS as a postdoctoral researcher, one of several new recruits. I had recently finished my doctoral studies in Australia and was excited by the prospect of working with like-minded people who had a growing reputation in artificial intelligence. As a result, I had a ringside view of the android project from its conception through to the very end.

When we crossed paths, Graesser would give me a friendly slap on the back, put his face uncomfortably close to mine, and bellow, “How are you doing, maestro?” I was flattered to be described as a maestro by such a respected researcher, until I realized that he greeted lots of people that way. “Maestro” was merely Graesser’s term of endearment for his many protégés.

The person most deserving of the endorsement was a new PhD student named Andrew Olney. Originally from Memphis, Olney had recently returned from Brighton, England, where he had completed a master’s degree in complex systems. Olney was wiry and sported a goatee and several facial piercings, including a metal stud in his tongue. His face seemed to alternate between two natural states: a pensive frown and an amused grin. He had returned to his hometown to be with his high school sweetheart, Rachel, now his fiancée.

Olney was bringing much-needed expertise to the back-end computational systems of the large projects at the IIS. He was also playing a key role with another newcomer, a psychology professor named Max Louwerse, in the language and dialogue group, one of the teams that were developing the AI that powered AutoTutor. But Graesser was already starting to think that Olney might soon need his own project. He was not sure what it might be, but it would have to be something that would really stretch the talents of this rising star.

By 2003 Graesser's summers had become annual pilgrimages across the conference circuit. He went to the meetings of the Cognitive Science Society and the Psychonomics Society, the two major conferences for any serious cognitive psychologist, and of Discourse Processes, a small society that focused on the study of natural language and in which he played a major role, and he attended a range of technology, engineering, education, and other conferences of varying interest to him, often with long, inscrutable acronyms, such as AAAI, SSSR, AIED, ICLS, AERA, FLAIRS.

That year, an old colleague in Santa Fe had asked him to come and speak at a small, invitation-only event, the Cognitive Systems Workshop. It was to be held at the Santa Fe Hilton and funded by Sandia National Laboratories, a quasi-governmental agency that explores cutting-edge technology. Graesser could hardly say no. Important people would be there, people who made decisions about federal grant money, as well as fellow researchers and even some former students.

Sitting in his office with wisps of cold winter air sneaking in through the window, Graesser flicked through a printout of the conference program to get a sense of which talks would be of interest to him. The presenters were, like him, mostly established academics with well-known research programs. There was, however, one exception: one presentation was to be given not by a tenured professor but by a graduate student from the University of Texas in Dallas. The title of his talk was wordy and cryptic: "Modeling Aesthetic Veracity in Humanoid Robots as a Tool for Understanding Social Cognition." At any rate, the conference was too small to have multiple streams of speakers running simultaneously, so Graesser would pretty much be obligated to sit through every presentation.

It could be good or it could be a waste of time. Either way, there were plenty of other smart people attending and they would have interesting things to say.

A thousand miles away, in Denver, David Hanson, the young man whose name Graesser's eyes had paused over, was becoming famous in certain circles. The annual conference of the American Association for the Advancement of Science—the AAAS—is a colossal event, with speakers and presenters from every conceivable branch of modern science and technology. The keynote talks provide insights into developments at the leading laboratories and universities, buffered by legions of presenters ranging from the eccentric to the pedestrian. Journalists wander among the drifting throngs, searching for something scientifically interesting but also catchy enough to appeal to a mass readership.

Hanson's presentation met those criteria. He was not a natural extrovert. In crowds he would be unassuming, blending into the background. But when he was at the front of a room with a microphone in his hand he was compelling. He talked about philosophy and art, society and science. His voice conveyed intensity and conviction, sometimes a hint of emotion, as he described his ideas about the future of humans and the technological world we are building.

Humans, he would explain, are creating a society that will be *more* than human. It will be a synthesis of human and machine, and we need to think about how to guide that process so that we create benevolent machines that enhance our experience of life. Hanson would move his arms around, emphasizing a point, and, occasionally, gaze at some unseen horizon beyond the walls surrounding them. He could engage people who had no expertise in science and, most important for the journalists, he usually brought a fully operational robot head with him to his talks.

Hanson was one of three speakers chairing a workshop at the AAAS called "Biologically Inspired Intelligent Robotics." The other speakers were Hanson's friend and mentor Yoseph Bar-Cohen, from the NASA Jet Propulsion Laboratory, a curly-haired rocket scientist who, despite his stature in the field, preferred to be known as "Yosi," and Cynthia Breazeal, a scientist from the Massachusetts Institute of Technology, in Cambridge. Breazeal, a senior researcher at MIT's Artificial Intelligence Lab, was interested in the effect of robot babies on human adults.

Hanson had originally planned to present his latest creation, K-Bot, but some last-minute adjustments had taken more time than planned. So instead he showed slides of his work and talked about his new invention: a synthetic substance he had started using for robotic skin that he called Frubber.

He told the audience about his first experiences building robots with wire frames and rubber skin. The problems with real rubber had quickly become apparent. For one thing, it is not very compressible. Rubber does not have the same folding properties as true skin. If a rubber mask is made with the same thickness as human skin, you need a lot of force to push it and pull it into different facial expressions. But if it is made much thinner, it does not look as lifelike and becomes weak and fragile. To make replicas of human faces that had the same emotional expressiveness as real humans he had to use large, heavy motors, which were difficult to work with. So he'd started experimenting with plastics and other materials. After a lot of trial and error he found a recipe for a compound that was much lighter than rubber, more pliant, and cheap to produce. It was also remarkably similar to human skin. He came up with the name Frubber and patented the formula.

At the end of his presentation Hanson apologized for not having his latest creation there for the audience, but he promised he would bring it to a special presentation the following day.

The next day, the room was packed with researchers, teachers, students, members of the public, and journalists. The chatter died as Hanson stepped forward. He held in his hands the disembodied head of a beautiful young woman.

"This is K-Bot," he said.

Hanson had modeled K-Bot on Kristen Nelson, a research assistant at the campus lab he worked at in Dallas and his girlfriend at the time. Under her Frubber skin, K-Bot had twenty-four motors that could pull her face into thousands of unique expressions, showing the full range of human emotions. She had cost just $400 in parts.

"In terms of complexity of the parts and expense incurred, K-Bot is not the most expensive in the world. But in terms of the sophistication of what it is capable of doing, it is the most advanced," Hanson told his audience. "It has the most expressive skin—it's a polymer developed in my laboratory—and has a compressibility comparable with human skin. It also has a high elongation, which means it stretches very easily."

K-Bot's "eyes" were mounted cameras connected to a computer that fed into a facial recognition program. The face "recog" could detect expressions on your face if you were looking at K-Bot, and K-Bot could mimic that expression right back at you: a kind of three-dimensional mirror that could only reflect emotions.

Hanson told the audience that he planned to make his robot more intelligent to provide a true interactive experience. He joked that the robot did not have a body and so, despite being modeled on a woman, technically could not be classified as male or female.

"I guess it is sort of an androgynoid," he said, and the crowd laughed.

That provided a catchy headline for the *Guardian* newspaper later in the week: "Human Face of Androgynoid." The subtitle that followed had a little more bite: "Nice Smile, but Shame About K-Bot's Personality." The article pinpointed a weakness in Hanson's work that he himself was aware of—the absence of intelligence. Certainly, he was developing a reputation for building sleek, attractive robots with unprecedented expressiveness, but beauty was not enough for him. After all, what is the point of a robot without a brain?

Hanson had trained as a sculptor at the Rhode Island School of Design, a small college in downtown Providence that sits between the Providence River and Brown University. The college has a small annual intake of students but receives thousands of applications and is arguably the most elite art school in the world, with alumni that include Seth MacFarlane, the creator of the animated series *Family Guy*; the New York architect Michael Gabellini; and David Byrne of the new wave rock group Talking Heads. The environment is conducive to experimenting with the synthesis of art and other fields, and Hanson dabbled with robotics at the college, once building a larger-than-life model of his own head that could be wheeled from room to room and have conversations with people through a remote device.

Even before college Hanson had developed an obsession with science fiction, reading the works of Isaac Asimov, Robert A. Heinlein, and, most of all, Philip K. Dick. Later, when he moved from art into robotics, his classical training in sculpture gave him the ability to create beauty where other roboticists produced mere functional mechanics. But such a background also left him at a disadvantage: he knew very little about AI.

After graduation, he moved to New Orleans and worked for Kern, a small sculpting company that had a steady stream of contract work for film companies such as Universal Studios and Walt Disney Imagineering. Two years later he moved to California to work at the Walt Disney Studios as a sculptor. The following year he moved laterally to a department known as Disney Technical Development. There he rediscovered his teenage sci-fi fantasies while working on several robotic projects.

Yoseph Bar-Cohen came across Hanson's work and invited him to make a lifelike robot head and give a small presentation to his colleagues at NASA. Hanson agreed and produced the first of many prototypes, a self-portrait with motors for facial muscles. The skin for that robot was standard urethane, and its expressiveness was limited. Bar-Cohen was enthusiastic nonetheless, and Hanson's work was warmly received at NASA. That gave him the confidence to build a second robot, learning from his mistakes with the first. He built a pirate head, complete with an eye patch and a leering grin. Pirate Robot's skin was made out of an early formulation of Frubber. Hanson's third robot was K-Bot. By this stage Hanson was beginning to refine his techniques. K-Bot was more advanced, had more motors, was more expressive and more physically appealing. K-Bot was also quicker to build.

Within days of the AAAS presentation, stories on Hanson and his work appeared in outlets ranging from *New Scientist* to the BBC. The *Guardian* was alone in its snarky reference to K-Bot's lack of brains. Other media reports emphasized K-Bot's impressive realism. Dan Ferber, in a long article on Hanson for *Popular Science* magazine, described the scene:

> Hanson, 33, walks in and sets something on a table. It's a backless head, bolted to a wooden platform, but it's got a face, a real face, with soft flesh-toned polymer skin and finely sculpted features and high cheekbones and big blue eyes. Hanson hooks it up to his laptop, fiddles with the wires. He's not saying much; it might be an awkward moment except for the fact that everyone else is too busy checking out the head to notice. Then Hanson taps a few keys and . . . it moves. It looks left and right. It smiles. It frowns, sneers, knits its brows anxiously.

Ferber declared: "K-Bot is a hit."

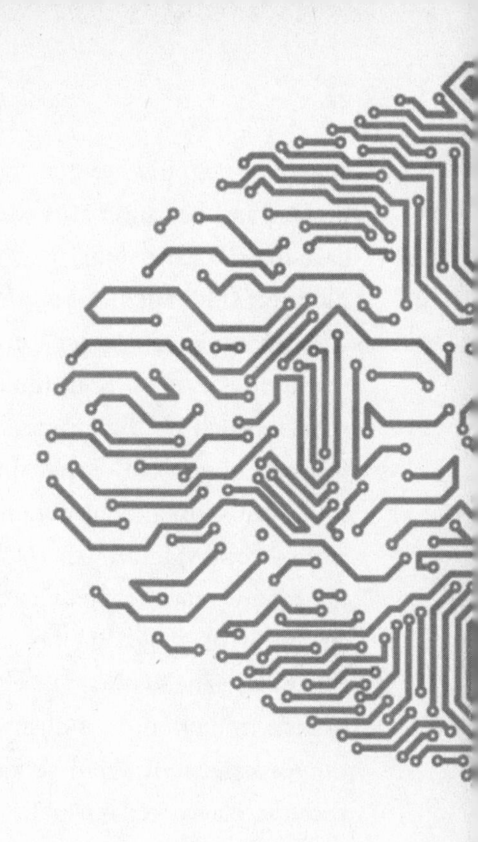

3. A Meeting of Minds

In 2003, the city of Santa Fe didn't have a major airport. Visitors had to fly to Albuquerque, an hour and a half away, then make their way northeast any way they could. The Santa Fe Hilton ran a shuttle bus between the hotel and the Albuquerque airport. Throughout the late afternoon and evening of June 29, the shuttle carried academics across the arid New Mexico plain to the hotel, where they clambered out with their suitcases and laptops, located their rooms, and then, if there was time, found each other for dinner or a drink.

The Cognitive Systems Workshop was an invitation-only event, a low-key, intimate gathering of around fifty scientists and engineers. They would spend the next three days in the Hilton's plush facilities talking shop.

Academic conferences are curious things. They exist so that researchers can transmit their findings and ideas more rapidly than through the byzantine, hidebound ritual of peer-reviewed journals. Feedback is instant and can occasionally result in flashes of spontaneous collaboration and insight. These meetings also provide a chance to network: ambitious graduate students and junior academics can brush shoulders with renowned figures in their field. But the glamorous locations and the opportunity for socializing mean that conferences are often also regarded as little more than an industry perk.

Graesser took them seriously. An extroverted and cheerful man, he considered one of the greatest advantages of conferences such as this to be the opportunity to hang out with like-minded people and catch up with old friends and acquaintances. Plus, and probably more important, he could efficiently advertise the achievements of his empire, the IIS. But having been on the summer conference circuit for over thirty years, he couldn't always maintain his enthusiasm. The Santa Fe Cognitive Systems Workshop was something of a reprieve for him. With such a small number of invited attendees, there would be a greater concentration of people he knew, fewer talks to attend, and fewer dud presentations.

The conference coordinator was Chris Forsythe, who years earlier had been one of Graesser's first doctoral students. These days he worked for Sandia National Laboratories. Sandia had given him the responsibility of bringing together some of the best researchers in the United States to talk about their latest findings and their ideas about developing advanced artificial intelligence systems.

Sandia has a long history of pioneering military research. Beginning with the Manhattan Project, it played a central role in the development of the nuclear arms race of the Cold War, before turning its attention to biological weapons analysis; it was involved in tracing the source of anthrax that was used in terrorist scares in 2001. The agency is perhaps best known in popular culture for its role in a mysterious plane crash near Roswell in 1947. Sandia's facilities are near the crash site known as Area 51, where UFOlogists insist an extraterrestrial spacecraft came to Earth. While Sandia was never implicated in the original accounts of the crash, conspiracy theorists have more recently suggested that there was a second crash and that the remnants of this crash, including six bodies, were taken into custody by Sandia representatives and transported to their laboratories. The U.S. government has investigated the rumors surrounding Roswell and found that these theories are incorrect and misguided.

Given its intense and long-standing involvement with the U.S. military research establishment, Sandia's interest in artificial intelligence was not just academic; its researchers were looking for promising leads for future military applications of artificial intelligence. AI had already made important contributions to military technology in areas as diverse as sonar systems, logistics and planning, and personnel training. In the 1980s, neural networks—artificially intelligent matrices that mimic the way a brain's neurons interact by being responsive to changing inputs—enabled a thirty-cent computer chip to detect underwater mines better than trained humans. Since then there had been an explosion of AI applications in weapons technology, including in guided missiles and aircraft control. Sandia wanted to determine where the next breakthroughs might occur, or where they were already occurring.

Graesser was known in military research circles because of his interest in artificial intelligence as well as in education, especially e-learning and computer training systems. He was a keynote speaker. On Tuesday, the second day of the workshop, he would give a detailed presentation on the greatest achievement of his lab to date, AutoTutor.

In a typical conversation with AutoTutor, a student sits at a computer, in front of an interface that includes AutoTutor's "face" and a dialogue box for talking to the program. At the time, AutoTutor communicated through Marco, an animated man with a large jaw and a serious, teacherly expression. As the image of Marco on the screen began to talk, his words also scrolled down in the dialogue box at the side, not entirely unlike an Internet chat session.

The ultimate goal was to turn AutoTutor into an agent that could teach any topic to a novice, on the fly. AutoTutor would "learn" a domain either by grabbing preformatted files online or through an easy-to-program interface that an expert could use to create an AutoTutor curriculum. During the workshop, AutoTutor was still in the prototype phase and was limited to discussing a handful of problems relating to conceptual physics.

The conversation might go something like this:

MARCO: Okay, let's begin. Hello, I'm Marco. I'm one of the computer tutors developed by the Tutoring Research Group, at the University of Memphis. I try to help students learn about physics by having a conversation with them. Let's get straight to the demonstration. Suppose a runner is running in a straight line at constant speed, and the runner throws a pumpkin straight up. Where will the pumpkin land? Explain.

STUDENT: The pumpkin lands behind the runner.

MARCO: Anything else?

STUDENT: I don't know.

MARCO: Okay. What is the horizontal velocity of the pumpkin?

STUDENT: The pumpkin has a constant horizontal velocity.

MARCO: So . . . the horizontal velocity of the pumpkin is the same as the . . . ?

STUDENT: Could you repeat that?

It was fair to say that Graesser was in love with AutoTutor, inasmuch as any man can be in love with a compelling idea or a magnificent work of art. But AutoTutor, while conceptually clever, suffered from the failings of so many university-born R&D projects. Its reason for being was not to fascinate or entertain or amuse but to demonstrate what could be done. It was a proof of concept, not a final product. The voice was tinny and monotonous, the vocabulary limited, the animated face bland and clunky. Marco looked in some ways like a pixelated zombie, grimacing in pleasure in response to correct answers and giving too-long frozen frowns in response to mistakes. Even so, AutoTutor was of interest to the small audience at the Santa Fe Hilton, a group of people with experience in looking past superficial characteristics and seeing the intricacies of the hidden architecture.

"Mixed initiative" was a key attribute of Graesser's system. In the study of dialogue and discourse, initiative means controlling the topic and direction of the conversation. If you start a conversation with someone, if you ask someone a question, or if you abruptly change the topic, then you have the initiative. A mixed initiative dialogue is one where either participant can take the initiative at any point.

Mixed initiative dialogue is familiar to humans, because in normal day-to-day conversations either person can interrupt, change the topic, ask a question, or go off on a tangent. But true mixed initiative conversation is hard for computers. AI systems either ask questions or answer questions. If they are "conversational" systems, they are usually limited to very prescriptive interactions that have been planned out by a programmer. It's risky to let a human take the lead because he or she can potentially divert the topic anywhere at all, creating unforeseen obstacles for the AI. At the time of the presentation in Santa Fe, AutoTutor's mixed initiative capability meant that it was at the cutting edge of current research.

Graesser also talked about the importance and the challenges of building artificial minds and speculated on new frontiers in the understanding of minds, language, computation, and social interaction. He is a polished performer and he was on familiar ground, so the audience was engaged, even if some people, researchers in similar fields who knew Graesser well, had heard much of it before. He wrapped up to polite applause, answered some questions, took a seat, and listened to the speakers who followed him.

Later in the lineup David Hanson, the graduate student from Dallas, appeared at the microphone. Above his face, sculpted and angular, his black hair was tousled, like a frieze of stormy ocean waves colliding with each other in all directions. A T-shirt hung limply from his shoulders, as if to convey his pointed refusal to dress for the occasion.

Hanson began by talking not about where the field of robotics was but where he thought it should be.

"Robotics," he said, "should be about more than functionality, especially as robots become increasingly ubiquitous. Instead, roboticists should be paying closer attention to how humans feel about robots."

He showed a series of photographs of his work in Dallas building robotic heads. It was clear that Hanson was dynamic, intelligent, and creative. Not only that, his robot was beautiful. This was unusual, given the setting: the workshop was more about ideas than surface appearances. Hanson's training meant he had the ability to create captivating and realistic three-dimensional machine faces. But, most important, he cared about how they looked.

It was also clear from his talk that Hanson could see the big picture. He mused about the changing relationship between humans and machines and what the future held for a society populated by increasingly intelligent machines. He explained how advances in machine technology meant we needed new ways of studying the way humans and machines were coevolving. It was a dazzling performance.

"I know talent when I see it," Graesser told me sometime later, "and I could see that he was a genius."

At the end of the session, as the room emptied and attendees made their way to the refreshment stand, Graesser and Hanson struck up a conversation. They were still talking over dinner that evening, and they started again the next day. They found common ground on all sorts of surprising topics, from robotics and AI to technology, science, art, and society.

I asked Hanson about the encounter. He said, "I went up to him and said that I thought his stuff was really cool, and then he told me he thought my stuff was really cool. And we went from there."

Graesser's expertise was software, while Hanson's was hardware; they had perfectly complementary skills. Graesser had done a lot of work with talking heads on computer screens: figuring out what people liked, what worked in a training environment, and how to make the programs seem "intelligent." He was intrigued by the idea of taking that work out of the computer monitor and putting it into physical, three-dimensional robotic heads. Here in front of him was a young man who might be able to make that vision a reality. He told Hanson that they should collaborate. What exactly the collaboration would involve was not clear, but they agreed that it would probably involve some kind of intelligent robot.

Graesser had achieved considerable success in finding support for developing education technology. There was a lot of grant money around for education research, and so ideas that had an education angle—such as e-learning—were easier to get funded. He suggested that this might be a productive avenue to explore, as a robot teacher could be a natural extension of AutoTutor. They decided to look around for grants and apply for funding later that year.

When Graesser got back to Memphis, he was immediately busy. The Institute for Intelligent Systems was expanding rapidly, with so many new people coming on board and so many new opportunities arising that he could barely keep up. His inbox overflowed every morning with messages from students, faculty, and collaborators from all over the world. Before heading to campus he would sip coffee in his home office as he worked through his daily correspondence.

AutoTutor had received a new round of funding from the Office of Naval Research. This grant would allow Graesser to incorporate three-dimensional simulations of physics problems into the system. The plan was to have AutoTutor talk to the student about the simulations as they were running, and even rewind or change the simulations to explain a concept, depending on where the conversation went. Graesser's team would also implement some more general upgrades to the interface to make it sleek and easy to use.

AutoTutor had been conceived more than fifteen years earlier. In that time, software interfaces had evolved from command-line DOS to Flash websites and multiplayer online video games. For a university group that specialized in prototypes, making your work look like the latest commercial application was a struggle.

Apart from the work with AutoTutor and 3-D simulations, Graesser was involved in several new projects, including Emotive Computing, a collaboration with MIT that aimed to create computers that could detect and react to human emotions.

There are many ways to get information about someone's emotional state. In the jargon of the lab workers at Memphis, these information sources are "channels." Graesser planned to develop a way of detecting real-time emotional states using several channels simultaneously. As someone sat using a computer, their emotional states would be recorded and integrated into the program they were using. The team planned to use AutoTutor to test the techniques, thereby killing two research birds with the same funding stone; Graesser would be able to use the Emotive Computing project to further enhance the capabilities of Auto-Tutor.

Some emotional channels are easy to read. A word someone uses, for example, can have explicit emotional content; "hate," "love," and "I'm angry" clearly indicate your emotional state to someone else. Other verbal indicators, like agreeableness ("yes" versus "no") or the general topic (fighting and conflict versus peace and harmony), can give more general indications. The force with which someone taps the keyboard is another channel. Hitting the keys hard might indicate agitation or aggression.

Posture is also a channel. Leaning forward indicates interest, whereas slouching indicates boredom. The team had a chair specially designed to detect posture, donated to the institute by a local company, Steelcase. The chair could produce a pressure map of its entire surface, showing what position someone was sitting in, how much he or she was moving around, and how much weight the person was putting on different parts of the seat. The students in the lab called it the "butt sensor."

Facial expression is a channel that most humans read instinctively and effortlessly. Even newborn babies can tell the difference between a happy and a sad face. By the time we grow into adults, we can usually parse the subtle differences between boredom and tiredness, resentment and shame, surprise and delight. Humans experience a vast array of emotions that are mixtures of a dozen or so basic emotions, and most of these play across our face to some degree: disappointment, wistfulness, consternation, resignation, anticipation, and thousands of others, even some for which there are no words or descriptors, but nonetheless we see, understand, and respond to all of them.

Getting a computer to read facial expressions is a far greater challenge.

MIT had recently developed an application, known as Blue Eyes, that could track facial features and determine what emotion was being displayed by them. To do this, a camera was mounted above a monitor facing a person. An image of the person's face was fed into the Blue Eyes program, which identified key features such as the eyes, eyebrows, and mouth and used this information to determine the expression on the face. Graesser's lab and MIT had struck up a research partnership to incorporate Blue Eyes into the Emotive Computing project.

The goal was to bring multiple channels together into an intelligent system that could monitor the emotions of a person sitting at a computer. With this information the computer could make an assessment about where to take a conversation. When the computer registered the emotion "boredom," it would respond accordingly, perhaps by changing the topic or suggesting a break. Frustration, happiness, surprise, interest, confusion: all these things could be determined, quantified, and used to modify the experience of interacting with an artificial agent on the computer.

There were other new projects at IIS. One was a study of how Internet users scanned a page of information on the screen when they were trying to learn about a topic. Max Louwerse was attempting to quantify gestural communication and look at ways that nonverbal information could be transmitted electronically. Another professor, Danielle McNamara, was collaborating with Louwerse and Graesser to develop an online tool for assessing the difficulty and complexity of texts. The tool, called Coh-Metrix, did not use the traditional readability formulas that focused on sentence length and word length but, instead, looked at the semantic structures of the text. This was a departure from mainstream methods of assessing books, and the researchers hoped that it would one day improve the grading and selection of school textbooks.

All this activity had brought a rush of money, which meant more people, more computers, and better, newer equipment. But with the new people, projects, and grants, the IIS had outgrown its current space. The solution was emerging across the street, behind the Department of Management. FedEx had recently made a large donation to the university in the form of a new building, and to show the university's appreciation for the company's corporate largesse, the building would be named the FedEx Institute of Technology.

FedEx's headquarters is near Memphis International Airport. For this reason, more freight moves through the Memphis airport on any given day than through any other airport in the world. Rumor has it that Memphis is so critical to FedEx's operations that in the event of a full-blown emergency, such as an earthquake or other natural disaster, the company has enough oil reserves and backup generators there to continue functioning at full capacity for three weeks. The FedEx Institute of Technology was never intended to be yet another functional stack of classrooms and tiny offices. FedEx's vision was to create a place where the best researchers on campus could be assembled, where they could collaborate with corporations, entrepreneurial ventures, and private organizations to do frontline research and produce new technologies and applications.

When I arrived, in January 2003, the site was a wasteland of rubble and tire tracks, fenced with chain-link. Signs read, DANGER! BUILDING ZONE. Six months later, the building was complete. As the structure neared its final stages of construction, its futuristic, four-story glass edifice towered above the aging buildings nearby. Each floor had meeting lounges with glass marker boards and panoramic views of the campus. In addition to the institute-wide wireless broadband service, there was high-speed Internet cable, with jacks and sockets fitted under floor panels throughout the building. The laboratory areas were spacious and light, with side rooms equipped with electronic whiteboards, retractable screens, and movable walls.

The university management had organized a committee to oversee the building—including the allocation of space—that would liaise with researchers and investors and put the necessary administrative infrastructure in place. One of Graesser's master's students, Eric Mathews, became involved, at first playing an unofficial role.

Mathews's slight, unassuming build and a quiet, disarming voice belie his analytical mind and talent for tough negotiation. He mounted the case for Graesser's lab to be housed in the FedEx Institute of Technology; the extent of his influence is unclear, but the IIS was allocated almost the entire fourth floor.

The move to the new building took two full days, using volunteer manpower from the lab's staff, students, and postdoctoral fellows. Everything that anyone in the IIS owned was dug out from the scattered dungeons of the psychology building: desks, cabinets, folders, files, computers, servers and server racks, tables, cardboard boxes full of old dot-matrix printouts, miscellaneous electronic cables and hardware, coffee cups. With no time to sort through the accumulated detritus of a long-standing lab and figure out what was useful and what could be thrown away, everything was moved.

During the shuffle Graesser was working on a major paper and putting together an application for a very large grant. He wanted to keep the relationship with Hanson going, but he did not have the time. Mathews had shown talent for networking, so Graesser asked him to keep in touch with the robot designer and try to make something happen.

Mathews said he would and started learning what he could about Hanson and his robots.

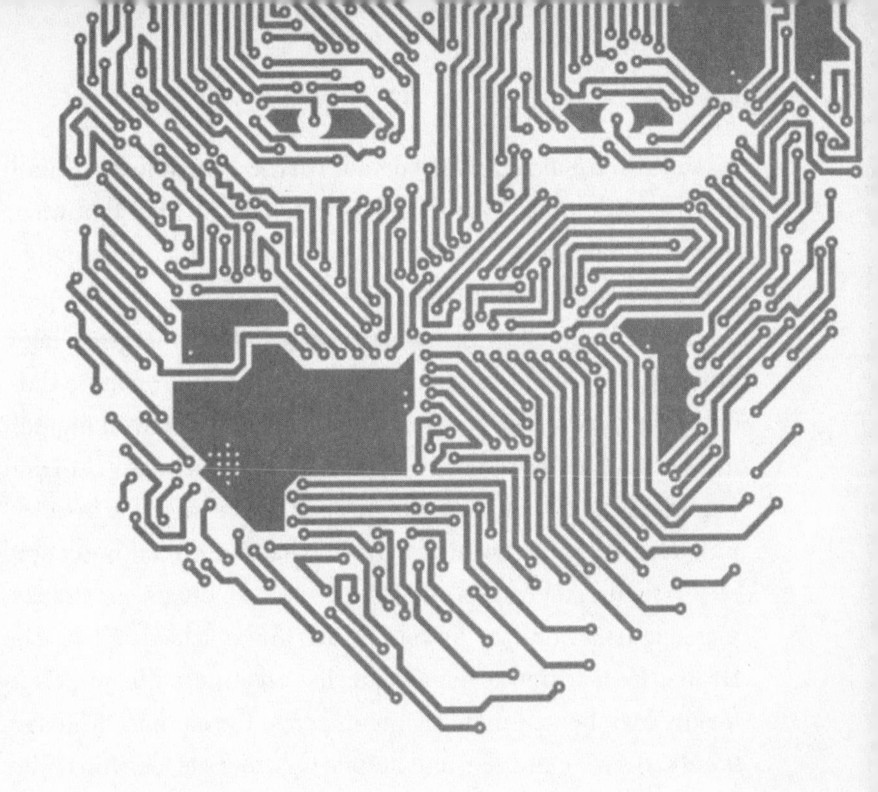

4. A Visit from Eva

Eric Mathews was something of an oddity as far as psychology graduate students go. While the other students were reading textbooks on therapy or semantic maps or, at least, Steven Pinker, he was reading *The Seven Habits of Highly Effective People*. In an ocean of jeans and self-consciously faded T-shirts, he wore khakis and collars. The truth was, he had discovered that he really wasn't that interested in psychology after all. He wanted to be a high-tech entrepreneur. Nonetheless, he was enrolled in the psychology graduate student program and had decided to make the most of it. He had already carved out a role for himself in Graesser's lab as the "go-to guy." He had learned that while university labs are often loaded with brains, there is usually a deficit of management skills.

Successful academics are typically put in charge of large budgets and expected to manage large teams of people. They often do okay with the budgets but are terrible with human resources. This was not the case with Graesser; he had been managing people for years and was pretty good at it. He believed in letting people grow and learn in whatever direction seemed to suit them, so he allowed Mathews to take on administrative, management, and liaising roles inside and outside the lab. The IIS had grown into a many-tentacled creature with numerous faculty members from departments across the campus. Management skills were needed, and despite Graesser's leadership ability the help was welcome.

Mathews soon got up to speed on David Hanson. They spoke on the phone a couple of times in the fall of 2003, and Hanson sent Mathews some press clippings and photos of his robots. Mathews quickly realized that Hanson's work was not just interesting, it was thrilling and unique. He also discovered that Hanson was basically a solo operator, working out of his apartment in Dallas with no funding or research grants. His only source of income was his student stipend. If there was to be any kind of collaboration, the initial financing would have to come from Memphis.

Mathews suggested to Graesser that they sponsor a visit from Hanson. It would be a relatively cheap investment and give them the opportunity to find out more about what he could do. Given that Hanson was a natural with the media, an exhibition of the

Hanson robots might also provide some publicity for both the FedEx Institute and the IIS. Hanson agreed to come, but because of the move and the launch of several new projects, as well as time constraints on Hanson's end, the event itself was put on hold until July 2004.

Graesser approved Mathews's suggested budget of $1,000 to cover airfare, accommodations, meals, and other sundry expenses for a week. Hanson would give a public demonstration of his robots in midweek. That would allow plenty of time for showing him off to the media and for trying to find some common ground that might lead to a fruitful endeavor.

Mathews asked Hanson whether he would be bringing K-Bot to Memphis. Hanson said no. He had built two more robots since K-Bot. The first of these was Vera. Hanson had continued to experiment with materials and had found a better formulation for Frubber, replacing a urethane component in the formula with silicone. Like K-Bot, Vera had also been modeled on his girlfriend at the time, although it was not the same girlfriend and her name was not Vera. Then came Eva. Though she was also made with silicone Frubber and was similar to Vera in many other respects, she was sleeker and more complex. Eva was to be his companion in Memphis.

Another person was brought into the loop for the Hanson visit: Andrew Olney, the chief programmer for AutoTutor. If they were going to turn Hanson's robots into conversationalists or do anything else related to AutoTutor, Olney needed to be on board. He was a proficient programmer in several computer languages, oversaw complex projects within the IIS, and taught computer science courses on campus. He had been impressed by the footage of K-Bot and was interested in meeting Hanson. Mathews and Graesser hoped that the two of them, Hanson and Olney, would strike up a rapport.

The physical contrast between Hanson, with his square jaw and muscular build, and Olney, the gangly beanpole with the pierced tongue, was stark. Yet they had a lot in common, and they knew that their job was to like each other and form some kind of partnership. The room that housed the Emotive Computing lab was empty that Monday afternoon, so Olney invited Hanson in there to talk. They checked out the butt sensor and chatted about Blue Eyes and MIT, AI in general, and the future of robotics. It was then that Hanson told Olney about one of his ideas. He wanted to make an android replica of a science-fiction writer, but he wasn't sure if it was possible.

Robots can be made in many shapes and sizes, to do many different things. They can operate surgical machinery, crawl across the ocean floor, build cars, and detect land mines. An android is a robot whose specific purpose is to look and behave like a human being. While robots are now commonplace, android robots are still mostly the province of science fiction. Olney asked which writer Hanson had in mind. Even as he asked, he had a hunch about what Hanson would say: Philip K. Dick. Olney thought it was a brilliant idea. They decided that they were the ideal team to do it, if such a thing could even be done.

The next day was the public exhibition of Hanson's work in the ground-level lobby of the FedEx Institute. The lobby is a vast, futuristic cavern with a Starbucks at one end and a stock market ticker tape at the other. Students lie around on brightly colored furniture, reviewing notes or tapping on laptops. That morning, Starbucks was doing brisk business as people arrived to see the presentation. An area in the middle of the lobby was partitioned off with metal stands and a knee-high rope. Inside the fence Olney and Hanson were getting ready.

The headliner of the day, Eva, sat on the table next to Hanson. Olney was going to demo AutoTutor using a speech-to-speech sequence. A student and programmer in Graesser's lab, Suresh Susarla, was rigging up microphones and helping to prepare the dialogue program. Soon a crowd gathered around the partition. A television crew turned up with a camera. A photographer started taking pictures. Andy Meyers, the executive director of the FedEx Institute, finally stepped forward and took the microphone.

"Thank you all for coming here this morning," Meyers said. A camera flashed from the crowd. "One of the goals of the FedEx Institute is to promote interdisciplinary research in cutting-edge technologies. So it is with great pleasure that I introduce a man who embodies that goal: Art Graesser."

Graesser accepted the mike. "In the past, right up to the present, if you want to learn something, you have to find a teacher who knows what you want to learn, and is available to teach you. Sometimes that's hard to do. Imagine being able to learn from an expert, anywhere, anytime. That's our goal with AutoTutor. We want to create systems that can capture the knowledge, the rare, valuable knowledge of true experts, so that thousands of people can benefit from that knowledge. So today, we're going to show you AutoTutor and we're also going to talk about the possibility of putting our programs into real robots. Andrew Olney will show you what is possible."

Susarla, running the software from a console to one side, initiated the program, and AutoTutor appeared on the screen. Olney leaned forward and spoke into the microphone.

"Hi," he said.

"Hello," said AutoTutor. Then they talked about AutoTutor's favorite topic: conceptual physics.

This real-time conversation was not a standard feature of AutoTutor. Typically, interaction was conducted by typing into a text box on the screen, but Olney had decided that the spoken word would be much more powerful for the public demonstration. He had been playing with implementation of speech-to-speech capability for the past few weeks. It was not perfect, which was why he alone would speak to AutoTutor that day. The speech recognition was calibrated to his voice; it tended to make mistakes when other people used it. A bigger problem was that members of the public were likely to say unpredictable things, which could, in turn, cause AutoTutor to say unpredictable things, or even crash.

After Olney came Hanson. He was holding what appeared to be a female human head. She had smooth skin, defined cheekbones, full lips, and brown eyes staring into the distance. She was beautiful.

"This is Eva." The cameras flashed.

"I was hoping to be able to show you some interactivity today, but at this time she has no responsive abilities," Hanson said. He explained her lifelike appearance and that she was also "capable of expressing the full range of human emotions." The software that controlled Eva's emotional expressions had not been set up in time. He could, however, demonstrate how she showed emotions. Cradling her head in one arm, he reached under her scalp, pulled slowly on something underneath, and Eva smiled. It was not a natural smile. It was more of a grimace. Her mouth twisted wide, while her eyes continued to stare out into space.

The problem, Hanson explained, was that facial expressions are the combination of many muscular movements, but he had made her activate just one set of muscles, widening her mouth. That is not how a real human face works. A realistic smile also involves upturned ends of the mouth and contractions near the eyes. Hanson reached under again, grimacing himself a little, and pulled on something else. The corners of Eva's mouth turned up, and the grimace was transformed into something that more closely resembled a real smile. It still did not look natural—her eyes maintained a vacant, staring expression—but it hinted that more was possible. It seemed as if Eva's humanity were there: close by, but out of reach.

Over lunch with Mathews and others from IIS, Hanson and Olney revealed their plan to create an android replica of Philip K. Dick. The questions flew. How, exactly, would Philip K. Dick's brain be represented? How would it interface with the robot? What would such an android do—just sit around and talk about science fiction? Or would it be more than that?

Then there were questions of intellectual property. Who owned the rights to Philip K. Dick's name and his physical appearance? What about his works? When the android spoke, it would no doubt be using Dick's body of work as a reference source, raising the issue of copyright.

There were questions, too, about how to frame the project. Was this an AI project? If so, what was the point of going to lots of trouble to build a realistic head? Was it a robotic project? Or perhaps art? How could it be described?

They decided that the solution was to view the android principally as an artistic exercise. It would be a robotic portrait of Philip K. Dick. Perhaps, if it was successful, robotic portraits would become popular for famous and historical figures. Rather than just looking at a painting or a photograph of someone, you could sit down and talk to him or her.

Olney thought he could cover the AI side of things. He would adapt the architecture he had used to build AutoTutor as a base for the conversational agent. As for the actual mind of Philip K. Dick, he could use a technique called latent semantic analysis, or LSA, to create a model of Dick's mental world. Mathews said he would look into the intellectual property rights. Naturally, Hanson would build the robot itself.

"What would we call it?" a student at the table asked. "Would it be called the Dick head?"

Laughter erupted. Finally, Mathews suggested they stick with the author's full name, just to be safe. It would be the Philip K. Dick android.

The next day, Hanson and Olney told Graesser about their new plan. It was clear from the blank look on Graesser's face that he had never heard of Philip K. Dick.

"Ever seen a movie called *Blade Runner*?" Olney asked him.

"No," said Graesser, but he continued to listen. Graesser liked what he heard, but he was skeptical about the proposal's ability to attract financing. He still believed that the best chance of securing money for collaboration with Hanson would be an educational project. He had even found a potential source of funding: the National Institute of Standards and Technology. NIST had a unit called the Advanced Technology Program that encouraged groundbreaking exploration rather than the common more "incremental" science, which works with established, well-worn methods and ideas. If ever there was an innovative project, this was it.

The proposal would be an expanded version of the idea Graesser and Hanson had discussed at the workshop in Santa Fe: to create a lifelike, intelligent robotic teacher, which they called an "autotutorial robot." It would combine robot heads created by Hanson and AutoTutor, or something similar to AutoTutor, developed at the University of Memphis. The team argued in their application that a robot would be a novel, engaging teaching device, able to hold a child's attention. In addition, the application proposed a series of studies into human-robot interaction, particularly in learning environments. The NIST application was their best chance at getting money, Graesser told Hanson. If they got funding from the Advanced Technology Program, then they could do the Philip K. Dick project on the side.

From Memphis, Hanson flew to New York, where he had scheduled a meeting with Chris Anderson, the editor of the technology magazine *Wired*, which had previously run articles on Hanson's work. *Wired* was also the principal sponsor of NextFest, an annual technology exhibition with a reputation for showcasing futuristic, eye-catching designs.

Hanson's head was full of his recent conversations in Memphis, and he was especially excited about the possibility of building a Philip K. Dick android. He mentioned the idea to Anderson and another of the magazine's senior editors. They asked for more details. Hanson gave them a bare-bones summary, no more than thirty seconds long. He told them that it would be a robot that looked like Philip K. Dick and it would converse as if it were actually Philip K. Dick, too. The editors said that the project sounded like just the sort of thing that NextFest needed, and that they wanted it for NextFest next year. Hanson accepted the invitation immediately.

That evening Anderson invited Hanson to dinner with editors from *Wired* as well as "some other people." Those other people turned out to be a Nobel laureate, various New York personalities, and David Byrne from Talking Heads. For Hanson, the conversation over dinner was "just mind-blowing."

The next day Hanson contacted Andrew Olney and told him that the Philip K. Dick android, when it was built, would be shown at the *Wired* NextFest in Chicago in the summer of 2005. At the end of his trip, Hanson packed Eva into a sports bag and carried her onto a plane headed from New York to Dallas. She was not even finished but she was already obsolete.

They had only just met, but Olney and Hanson already shared many memories, as they had both been obsessed with science fiction as teenagers. They both particularly loved the stories and novels of Isaac Asimov and Philip K. Dick.

The prolific Asimov was the better known of the two during their youth, having written the acclaimed Elijah Baley trilogy, about a detective who teams up with a robot to solve murders, as well as several short-story collections, such as *I, Robot* and *The Rest of the Robots*. Across more than seventy books, he set his stories anywhere from the late twentieth century to millions of years into the future. In his 1942 story "Runaround," he proposed the "Three Laws of Robotics":

1. A robot may not injure a human being or, through inaction, allow a human being to come to harm.
2. A robot must obey any orders given to it by human beings, except where such orders would conflict with the First Law.

3. A robot must protect its own existence as long as such protection does not conflict with the First or Second Law.

Asimov's rational robots and the three laws that guided their creation provide a fanciful peek at the dilemmas that will face humans in the future. If we build robots, how do we make sure they work for us and not against us? How can we make sure that they don't destroy humanity? His laws seemed to provide the solution. Following them, Asimov's characters mapped out a path of ever-expanding human civilization across the galaxy. This vision of benign robots that aid humankind in our conquest of Earth and the stars captivated Hanson.

Philip K. Dick, on the other hand, imagined all the ways robots could do us in. Like Asimov, he loved androids, but he was also afraid of them. Though none were built during his lifetime, they filled his head and populated his stories. That androids would one day walk the Earth, he was sure, but he grappled with their nature. Would they be good or evil? Would they think they were humans? Would they take over the world? Would they be the next step in evolution?

Science fiction had a dedicated but small following in the 1950s and '60s, so writing it did not pay well. In order to earn a living from science fiction, Dick became astonishingly prolific. Between his first published work, in 1952, the short story "Beyond Lies the Wub," and his death, in 1982, he wrote and had published forty-eight novels and countless short stories. In 1964 he wrote six books, all of which were commercially successful. His typical settings were an authoritarian police state, a society controlled by megacorporations, or a postapocalyptic wasteland. Shortly before his death he told a friend he had reached the point where he could produce a science-fiction novel in a week. Unfortunately, Dick's copious output was not entirely the fruit of his considerable natural ability; amphetamines also helped.

Several of Dick's books are now considered classics, but he is best known for his 1968 novel *Do Androids Dream of Electric Sheep?*, a thriller set in a dystopian future where Earth is terminally poisoned by war and everyone who could has fled to other planets. It is the story of a policeman named Rick Deckard who must find and destroy some renegade androids. Deckard's job is complicated by the fact that these rogues are the very latest model, Nexus-6, and almost indistinguishable from humans. They bleed like humans, because they are constructed from organic components. They're as smart as humans, because they have been given advanced artificial intelligence. The only thing they seem to lack is human empathy. Deckard must hunt them down and kill them, though "kill" is not quite the right word, since the androids aren't technically alive. Deckard and the other police refer to it as "retiring" the androids.

Deckard's challenge is to see beyond the surface of the people he meets and figure out who or what is underneath. It is a problem faced by many characters in Dick's work. In the short story "The Father Thing," a boy learns that his father's body has been taken over by an alien, and kills him. In "Second Variety," androids working in factories during World War III have been programmed to evolve and are totally autonomous. They have begun manufacturing android soldiers, who appear to be refugees or small children but, once they have made their way inside the enemy's defenses, turn into killing machines. Four soldiers sit in a bunker, aware that one of them must be a killer android, yet not knowing which one is. These characters, by Dickian standards, get off lightly.

Another common fate is for his characters to discover that their identity or their entire universe is an illusion. For some this means learning that they are not human, as they previously believed, but an android whose memory has been programmed by a robot-manufacturing corporation; for others, that they are not, after all, a soldier in a platoon stationed at a jungle outpost but a member of a band of paranoid schizophrenics who escaped from a prison for the insane.

In 1963 Dick received the Hugo Award, the most prestigious honor in science fiction, for *The Man in the High Castle*, a novel set in an alternate universe where Germany and Japan won World War II. Whereas many of Dick's books are heavily plotted page-turners, *The Man in the High Castle* is slow and whimsical. It does, however, include his trademark reality twist. Characters move backward and forward between the world of the novel and the world of the reader, with some help from the Chinese fortune-telling system explicated in the *I Ching*. The man of the title is an author writing his own counterfactual novel, in which Germany and Japan lost the war. Dick also won the John W. Campbell Memorial Award and was a Hugo Award nominee for his 1974 novel *Flow My Tears, the Policeman Said*, about a man who awakens in a parallel world with no identity. These accolades represented approval from the science-fiction community, where he was already well known, but they did not make him famous.

Several years after *Do Androids Dream of Electric Sheep?* was published, Warner Brothers bought the film rights. In 1980, Ridley Scott was contracted as the director and Harrison Ford was hired to play Deckard, with Rutger Hauer tapped to be the leader of the rogue androids. The studio people thought the title was not catchy enough, so they called it *Blade Runner.*

As is always the case, the movie is rather different from its source material. Deckard's wife is out of the picture, and there are no electric sheep living on his roof. There are also changes in the nature of the androids. In the novel they have no sense of compassion for the living, and in fact hate and resent humans; in the film the line between human and robot is more blurred. The calculating, emotionless human predator Deckard is contrasted with the runaway robots, who are on a futile mission to change their destiny. Although he took liberties with Dick's work, Scott managed to capture the dismal mood of a futuristic society in decay. What the movie has in common with the book is a haunting, sad picture of what Earth could become, a complex exploration of what it means to be human, and one man's conflicted mission to kill.

Dick died of a series of strokes in March 1982, only weeks before the public release of *Blade Runner.* His death has been attributed to the drugs he relied on to maintain his productivity. Dick himself knew that his drug use might come at a cost. In an interview recorded by his friend Gwen Lee two months before Dick's death (and later published in the book *What If Our World Is Their Heaven?*), Dick had predicted, "Finally the cost is going to be higher than the yield line."

Before he died, Dick saw a preview of *Blade Runner* at a private studio screening. There are conflicting versions of his reaction. He reportedly had a heated argument with Ridley Scott, but they settled their differences and Dick endorsed the film. He was particularly impressed by the huge police building that towers over the city at the beginning of the movie. He said that it was "absolutely my fantasy of what it would have to be like forty years from now."

Blade Runner did not perform well at the box office, failing to sell enough tickets to recoup its production costs. But it proved successful on the video market, and within a few years it had gained a reputation as a cult classic. In 1992, a director's cut was released into theaters, with a "final cut" version following in 2007, for the twenty-fifth anniversary of the movie.

A second Dick story, "We Can Remember It for You Wholesale," reached the big screen in 1990. In the future depicted in the story, there is no need to go on expensive and troublesome vacations; scientists can instead insert a chip into your brain with the memory of having gone on a trip. All you need to do is visit a clinic and have the memory implanted. After all, why do people go on vacation except for the memories?

The story takes this premise to the next logical step: if you can have a memory implanted for any vacation at all, there's no reason to settle for a boring beach holiday or a weekend at a campground. Why not be a secret agent or do something else equally exciting? The catch is that someone who really is a secret agent, but whose memories have been suppressed so that he can be a convincing "sleeper" in the humdrum of suburban life, walks into such a clinic. The false memories inserted by the clinic, called Rekall, clash with his true memories, blowing his cover. As before, Dick's cumbersome title was discarded for something catchier: *Total Recall.* Unlike *Blade Runner,* it was an instant hit. In all, ten Dick stories have now been made into films, and counting.

In the meantime, Dick himself has become something of a cult figure, in part because of the growing reputation of *Blade*

Runner, in part because of the unique space his books occupy in the literary world. If you want dystopia, police states, and reality warps with paperback sci-fi action thrown into the mix, Dick's your guy. His reputation as an author who lived with mental illness and whose work was drug-inspired also fuels his fans' admiration—that this combination of chemicals, insanity, and pop culture combined to produce brilliant art is at least partly true. Dick has even been credited by some as being the driving force behind the resurgence of interest in Gnosticism.

While the work of other science-fiction writers of similar standing has disappeared from the shelves of secondhand stores, Dick's books, even those that were out of print a decade ago, are today mostly in circulation and being reprinted. The Philip K. Dick Award is now a prestigious honor given annually to outstanding new science-fiction books. Since his death Dick has emerged as one of the most significant writers of the twentieth century.

Hanson felt he had a personal connection with Dick. For a teenager, Dick's androids were more compelling, more empathic and tortured, and therefore more human than Asimov's, for whom characterization was never a strength. Indeed, Asimov's humans are sometimes so devoid of emotions that they could be mistaken for his robots. At the same time, Hanson identified more with Asimov's optimism about the prospects for humanity's relationship with androids than with Dick's bleak negativity.

While the young Hanson had dwelt in the robot worlds of a small number of authors, Olney had been a sci-fi omnivore. His reading started in his teens with the purchase of a space opera from the local newsstand, then moved into the science-fiction section of a nearby bookstore. In the space of a few years his collection of sci-fi paperbacks grew until it was, as he described it, colossal.

Olney told me about his addiction one night at a Halloween party at his house near campus. It was a mild, clear night, perfectly suited to a backyard party. We sat under the burning wicker lamps while students, mostly from Graesser's lab, stood about in outlandish costumes and drank keg beer.

Being familiar with Asimov, I asked him if he had read the Foundation series.

"No, I never got much into Asimov," he said. "I liked Poul Anderson and Frederik Pohl."

"What about Philip K. Dick?"

"Oh, yes. He's good."

We fired names of authors and book titles back and forth, testing for common ground: Robert Silverberg, Brian Aldiss, Frank Herbert, *Ender's Game, Neuromancer, The Chrysalids.* His breadth of reading and his knowledge of science fiction were both vast, and I could not keep up.

Olney explained that for him science fiction had become more than just an enjoyable escape. He had stopped reading it after high school, but the habit returned while he was studying for his master's in Brighton, in southeast England. Unlike car-oriented Memphis, Brighton was easy to get around by foot; he walked everywhere. Every day en route to campus he passed three used-book stores, each of which had an incredible catalog of science-fiction paperbacks, novels, and short-story collections from the 1950s and '60s, the so-called golden age of science fiction. He found classics, out-of-print works by the genre's most creative minds, going for as little as fifty pence—the equivalent of less than a dollar.

As his personal library grew, his desire to read sci-fi became almost insatiable. He would look forward to getting home so that he could retreat to his room and become immersed. He would even reread novels he had read several times before. Reading science fiction consumed all of his spare time. He eventually realized that he had a problem. The only way to break the spell it held over him was to throw all of the books away, and that is what he did. He threw every last science-fiction book—several boxes of them—into the trash. He has not read a science-fiction book since.

"Nothing?" I asked. He shook his head.

I asked him about television and movies. Had he banned them, too? Did he avoid turning on the Syfy channel?

He laughed. "No," he said. "TV was never the problem. It was books."

With his science-fiction books gone, Olney needed to find other ways of escaping from the banality of the everyday world. He channeled that sense of wonder at the possibilities of new technology into his studies. After all, he told himself, computer science is kind of like science fiction without the fiction, creating complex new worlds and pushing the boundaries of human technology—in the real world, rather than on the page.

5. Simple Robots

Hanson's visit to Memphis with Eva had been a success. Friendships were made, ideas were born, and grant applications were now in the pipeline. A second visit was planned for the late fall, to coincide with the first anniversary of the opening of the FedEx Institute of Technology. The initial joint exhibition had showcased both Memphis, with AutoTutor, and Dallas, with Hanson's Eva. That was good, but Olney and Hanson hoped to take the next display to a new level by incorporating artificial intelligence into Hanson's robots. For this to happen, Olney, an AI expert, would have to learn a few things about robotics.

The relationship between robotics and AI is the same as the relationship between hardware and software. The physical parts of a computer are the computer's hardware. This includes the screen and keyboard as well as the tiny wires and circuits inside, the hard drive, the motherboard, a plastic sheet with basic wiring connecting the computer's interior parts. The software is the stuff that runs on the computer, like Windows and Google and computer games. Engineers make hardware; programmers make software. A microwave is hardware. The microwave's timer, software. A TV is hardware. A TV remote is also hardware. When the remote switches on the TV, that's software. The distinction between hardware and software permeates all modern technology.

Engineers know, however, that there are areas where the distinction isn't clear. For example, a computer's motherboard is run by a small chip called a BIOS that has its programming hardwired in. The BIOS is both hardware and software. In robotics, too, the difference is often difficult to see. If an engineer wires a circuit just right so that a robot will recoil from sharp objects or beep when it detects a mine, is that hardware or software? And what about a thermostat? It's a kind of program, but the program is embedded in the machine itself.

Olney was a software guy. In his day-to-day work he programmed computers and made them do things. Certainly on the AutoTutor project this involved some tinkering with hardware. Now he had to learn to program robots. Hanson knew quite a lot about this, but Hanson lived in Texas and they were both busy, so Olney pretty much had to tackle it on his own. He was not even sure of the scope of the task he had given himself. How tricky would it be to get a robot to talk?

He found a website that gave instructions on how to turn standard electronic toys into robots. Any battery-powered toy has some kind of circuitry, and with the right equipment that circuitry can be modified. Olney remembered a novelty Christmas present, a singing fish, that his mother had given him a few years earlier. The rubber fish was mounted on a board like a fisherman's trophy, with a plaque underneath that read, BIG MOUTH BILLY BASS. If someone clapped or walked past, Billy Bass would sing a song. One of the songs in its repertoire was "Take Me to the River" by Talking Heads. The fish would flap its tail in time to the music and its mouth would open and close as if it were singing the song.

Sure, it was funny, but as with all novelty presents the joke faded fast. Billy Bass had spent the last few years in a box of junk in the corner of Olney's garage. He pulled it out of the box, dusted it off, and took it inside. The batteries were puffy white from acid rot. He cleaned it up, installed new batteries, and switched it on. It still worked.

As hardware went, it was simple, just a rubber fish with a speaker and a movement detector, but with the right modifications it could probably run some sophisticated software. Olney thought he could get it to run AutoTutor. If he could, that would prove that AutoTutor could, in principle, be run on serious hardware like Hanson's robots.

He peeled off the rubber skin and studied Billy Bass's insides. They were surprisingly well designed and, for a toy, quite complex. They included a photoreceptor that could detect nearby movement. The photoreceptor was connected to a switch that activated the song-and-dance routine, which was how Billy Bass knew when to start singing. The photoreceptor could be easily bypassed. Olney wanted Billy to talk in response to verbal input, not because someone jumped or waved their arm.

Underneath the rubber skin was a crude skeletal structure. It had three distinct parts, each with its own motor: the head, the tail, and the body. The three motors and the speaker all had wires running from them to a circuit board with a control chip. That chip ran the whole show. It was the fish's artificial brain.

From Olney's point of view, the control chip was a lost cause. He could never reprogram that. For one thing, it was encased in a tiny box. Even if he opened it without breaking it, there would be little he could do. The box and chip had been equipped with specialized hardware and software either at the original factory or during some subsidiary operation, and reprogramming would probably require microscopic adjustments to various internal switches.

A simpler approach would be to reroute the wires on the circuit board. The circuits that connected the motors and the speaker to the control box could be diverted along new paths to a control box of his own making. He could even send the circuits to a computer that had AutoTutor installed. This was easier because the circuit board wasn't hard to figure out or work with. It was large as circuit boards go, so he could tamper with the wiring with minimal risk of damaging something. Each circuit running into the control box was connected to a single component of the fish. One opened the mouth, one closed the mouth, one operated the speaker, and so on. Each one was just an on-off switch. Even a novice could tinker with a circuit as simple as the one inside Billy Bass.

Olney purchased a programmable board from a local electronics shop and rerouted the circuits into it. By changing switches on the board he could make Billy Bass do his bidding. The first stage of his mission was accomplished: he controlled Billy Bass's movements.

The next stage involved the speaker circuit.

Billy Bass's electronics featured a small memory load of prerecorded music that fed straight into its speaker, but that was easily diverted. He could connect the speaker circuit to the speech synthesizer on his laptop and output human speech instead of the song. That also meant he could run AutoTutor through the fish. When the program had something to say, the fish could say it.

For this to work, Olney had to tweak AutoTutor. Outputs that were sent to the screen had to be sent to new modules and, ultimately, to the fish controller Olney had built. The messages had to be commands the fish controller would understand, so the complex version of AutoTutor that currently ran on PCs would not work on Billy Bass. He had to simplify or eliminate many elements of the program. For example, sometimes Marco, the animated teacher of AutoTutor, would move his hand in a circular motion, encouraging students to respond if they were taking a long time. This was changed to a fish's tail flap. Speech was rerouted to the fish speaker. Lip movements were simplified to the binary open/shut of the fish's mouth and synchronized with the beginnings and endings of each speech act.

After several weekends of tinkering, AutoTutor was running through Billy Bass. Instead of entertaining passersby with renditions of a new wave hit, Billy Bass lectured them on Newtonian physics. Olney filmed Billy Bass giving a lesson on gravity. In the footage, the fish says, "Suppose a runner is running in a straight line at constant speed and the runner throws a pumpkin straight up. Where will the pumpkin land? Explain."

The fish then flaps its tail impatiently, like a schoolmaster tapping a desk. No answer is forthcoming, so eventually the fish answers its own question: "The horizontal velocities are the same."

The Billy Bass experiment was a success. The next step would be the real thing: implementing AutoTutor or some other complex AI system on Hanson's robots.

Hanson didn't want to make a mere talking machine. He wanted to make something that would really seem to be alive. Visually, he knew he could pull it off. He was a classically trained sculptor and had crafted several previous robot heads. He knew he could make a head that would look just like the middle-aged science-fiction writer. And Olney would add an extra dimension—things that would help make it not only look like the dead author but behave like him, too. Hanson's earlier robots had captivated audiences and scientists with their humanlike qualities and lifelike emotional expressions. This one would be imbued with the same physical beauty and also have a voice. It would be the most lifelike of all.

Olney, despite his success with Billy Bass, still had significant concerns. As far as he could tell, nothing exactly like this had been done before, though similar projects were being developed at Ivy League universities by huge teams of technicians with budgets of many millions of dollars. He and Graesser had developed conversational agents on computers; he knew that getting computers to talk and sound intelligent was not easy, and the current technology was limited. As for creating a convincing simulation of a human, in AI circles the ultimate goal had been set: getting a machine to pass the Turing Test for intelligence.

The Turing Test had been conceived by Alan Turing, a British mathematician who worked as a code cracker during World War II. The Germans had invented and built a secret device, known as the Enigma machine, that could generate an infinite number of codes. In fact, it could generate a new code every day. Standard code cracking did not work against it. The secret lay in figuring out the mechanics of the machine generating the codes. Allied code crackers did not have such a machine, and they did not know how one was built.

A team of cryptographers spent most of the war working on this problem, and they did eventually crack the Enigma machine code. It was an extraordinary achievement of problem solving, perseverance, and mathematical brilliance. Their success came in time for the Normandy D-Day landings and provided valuable intelligence on German troop positions and panzer tank movements. During the invasion, the Allies intercepted high-level messages between German commanders and, thanks to the cryptologists, knew exactly what the Germans were thinking and how they planned to respond.

The code-breaking headquarters in England was housed in a large country manor called Bletchley Park, which contained a radio intercept room known as Station X. The Station X team included the brightest mathematicians of their generation, but to succeed they had to move beyond paper-and-pencil solutions and invent new kinds of problems and methods. Their work led to a clearer understanding of the difference between the machines that generated the information (the hardware) and the patterns themselves (the software) and laid the foundations of modern computing theory.

One of the leading cryptographers, Alan Turing, came up with the idea of building physical simulations of what the Enigma machine was doing. These constructions were known as the "bombe." Turing's ingenuity helped solve one of the most diabolical puzzles in mathematical history. It also forced him to think about codes, and information in general, in new and unusual ways.

After the war, Turing continued to explore theories of information. He described, in mathematical terms, how computers should work and so made today's ubiquitous desktops and laptops and handhelds possible. Turing started by imagining machines that could carry out a series of commands, much like the Enigma machine generating a new code for the German generals. Other machines could execute different commands, like counting objects or sorting a pile of blocks from largest to smallest. These commands could be programmed in advance using a ticker tape with carefully placed holes. The pattern of the holes would tell the machine what to do next—an abstract version of a computer program. Turing described how such a machine—now called a Turing machine—could be built to follow any conceivable series of instructions, given the right instruction tape.

Turing's two seminal achievements—cryptography during World War II and computing theory afterward—secured his place in history as a great mathematician and logician. Despite these accomplishments, he is probably most famous for a particular paper he wrote in 1950, called "Computing Machinery and Intelligence." At the time, barely a handful of computers existed and the general public was, at best, only dimly aware of them. Those few computers were enormous machines stowed in secure military establishments; they did not have the versatility or elegance of modern computers and were mostly used for performing complex or tedious mathematical calculations. Turing saw beyond the valves and clunky machinery and realized the possibilities that computers held. With the right program they could do anything, even think like a human. Perhaps they could even think "better" than us. If that happens, Turing wondered, how will we tell human from machine?

The starting point of his paper was the question "Can a machine think?" While this was an intriguing topic, Turing thought words like "machine" and "think" were simply too ambiguous to be useful. He suggested that a better way to solve the problem was to invent a test for intelligence—and, of course, he had such a test in mind. Viewing it as a sort of parlor entertainment, he called it the "Imitation Game." These days it is called the Turing Test.

Imagine that you have a computer and you have programmed it to act like a human. The computer can hold a conversation like the following, written by Turing as an example of what might be possible.

> QUESTION: Please write me a sonnet on the subject of the Fourth Bridge.
>
> ANSWER: Count me out on this one. I never could write poetry.
>
> QUESTION: Add 34957 to 70764.
>
> ANSWER: (*Pause about 30 seconds and then give as answer*) 105621.

Question: Do you play chess?

Answer: Yes.

Question: I have K at my K1, and no other pieces. You have only K at K6 and R at R1. It is your move. What do you play?

Answer: (*After a pause of 15 seconds*) R to R8 mate.

Such a performance seems intelligent. The task is to find out if the computer is, in fact, intelligent. Here is how the game is played: A human is brought into the testing room as the interrogator. In a second room, hidden from the interrogator, are the computer and a human. The interrogator does not know who is the human and who is the computer. His or her job is to figure out which is which by interrogating and talking to both.

In Turing's day, the output of a computer was always printed on punched paper, so to make things fair for the computer all conversations between the rooms had to be typewritten. If the interrogator converses with both and can't tell which is human, then the computer has passed the test. It is intelligent.

There has since been half a century of debate, argument, and speculation about the Turing Test. Is it the right way to tell if a computer is intelligent? And will a computer ever pass it? The most controversial aspect of the test is that it is hardware-independent. Turing was not concerned about what kind of machine the interrogator was talking to. It did not matter if the computer was made of metal or plastic or wood, powered by electricity, steam, or some other source. The only thing that mattered, and the only criterion for intelligence, was the output.

Therefore, Turing saw no point in making machines that looked like people, even if it could be done, since the result would be little more than decoration. Appearances weren't the main game; what went on inside was what counted. He wrote, "No engineer or chemist claims to be able to produce a material which is indistinguishable from the human skin. It is possible that at some time this might be done, but even supposing this invention [becomes] available we should feel there was little point in trying to make a 'thinking machine' more human by dressing it up in such artificial flesh."

Hanson was a huge admirer of Turing but thought he was wrong on this point. Surely part of being human is the feel of your skin, the twinkle of your eye? Could it ever merely boil down to being able to add large numbers together or come up with a winning chess move? Such clever responses might be all that was needed for a wonk who spent his life in secret military labs and dusty university offices, but Hanson didn't think that was a realistic test of being human, or even the full range of what it meant to be intelligent. For Hanson, the core idea of the Turing Test was that it endorsed the goal of making machines that acted like humans. The very existence of the test itself implied that this was a difficult but worthwhile endeavor.

Philip K. Dick also knew of the Turing Test. He admired Turing's work but did not believe the test had merit. For one thing, Dick thought that the robots of the future would easily pass the test, so the exercise seemed doomed to be a historical curiosity. He believed that superintelligent computers were around the corner and would be able to pass any intelligence test, Turing or otherwise, we could throw at them. Dick's stories were populated with robots posing successfully as humans, even becoming uncertain about whether they might actually be human, as well as humans who appeared to be robot impostors. He was sure that the line between human and robot would eventually become blurred beyond recognition.

For Dick, the biggest problem with the Turing Test was that it placed too much emphasis on intelligence. Dick believed that empathy was more central to being human than intelligence, and the Turing Test did not measure empathy.

In *Do Androids Dream of Electric Sheep?* Deckard, hunting rogue androids, is supplied by his superiors with a setup for telling humans and androids apart, called the Voigt-Kampff test. The test consists of a series of questions and a device for monitoring the responses of the individual being tested. The Voigt-Kampff questions are designed to provoke an emotional reaction in humans. A human will have a physiological response to the question within a very small and precise window of time—say, pupil dilation or a facial blush occurring in a fraction of a second. The machine detects these responses and the time it takes for them to appear. The theory is that androids, while able to pass themselves off as human in most situations, have trouble mimicking emotions in real time.

In the novel, Deckard administers the test to Rachael Rosen. Rachael is the niece of Eldon Rosen, a wealthy businessman and the founder of the Rosen Association, the company that makes androids. Deckard is told that she will take the test so that he can see how a human responds. In fact, it is a trap set by Rosen's corporation to try to discredit the test. What Rosen does not tell Rick Deckard, and what Rachael herself does not know, is that she is not human but an android.

> After making a jot of notation Rick continued, turning to the eighth question of the Voigt-Kampff profile scale. "You

> have a little boy and he shows you his butterfly collection, including his killing jar."
>
> "I'd take him to the doctor." Rachael's voice was low but firm. Again the twin gauges registered, but this time not so far. He made a note of that, too.
>
> "You're sitting watching TV," he continued, "and suddenly you discover a wasp crawling on your wrist."
>
> Rachael said, "I'd kill it." The gauges, this time, registered almost nothing: only a feeble and momentary tremor. He noted that and hunted cautiously for the next question.

Rachael fails the test. She always believed that she was human, but after taking the test she learns that she is an android, that her memories of her human childhood were merely wired onto a chip, programmed into her to create the illusion of humanity.

Just as the Turing Test was designed to distinguish dumb machines from smart machines, the Voigt-Kampff test that Dick made up was used to distinguish smart machines from humans. In a leap of imagination, Turing had anticipated that one day we might have trouble assessing the intelligence level of machines. Dick went further: in addition to assuming that machines would be built that would pass the Turing Test, he assumed that engineering would advance to the point where machines not only were as smart as us but would be physically indistinguishable from us, too. Deckard is not equipped with a Turing Test, because in the future portrayed in *Do Androids Dream of Electric Sheep?* such a test would be useless. For this reason, Deckard probes his subject not for intelligence but for empathy. The androids he is hunting lack humanity, in the old-fashioned sense of the word.

This is where Olney and Hanson differed in outlook. Olney spent his evenings trying to figure out how to make androids intelligent. Hanson wanted to make androids human, which was an entirely different thing.

6. The Artist as Scientist

Hanson returned to Memphis in October to speak at the first anniversary of the opening of the FedEx Institute of Technology. On the previous visit, he had come alone. This time, he brought a small entourage. Derek Hammons was a programmer who had helped Hanson with earlier robots, writing code to make them perform rudimentary actions, such as showing various emotions and looking around; Steve Prilliman was a marketer and long-standing friend of Hanson's.

When they got to campus, Hanson's team from Dallas met

the team from Memphis: Mathews, the networker and liaison between Dallas and Memphis; Olney, the programmer; Graesser, the director of the IIS; and Suresh Susarla, the graduate student in computer science. They talked about the NIST Advanced Technology Program application that was due soon. Everyone agreed that they had a strong chance of winning the grant. When it came through, they would be able to start work on building the Robot Teacher and, as agreed previously, the Philip K. Dick android would be piggybacked onto the project's resources and infrastructure.

Hanson introduced the programmers, Hammons and Olney, to each other. After the meeting they installed Eva on a table in a lab room on the fourth floor of the FedEx Institute. Hammons pulled his code up onto the screen and talked Olney through it, explaining at each step what he had done and why. They ran the code, and Hammons, sitting at the computer, made Eva smile, frown, and turn her head from side to side. That evening Olney looked through Hammons's code on his own. It was certainly useful. Hammons knew Hanson's machinery well and had solved several problems that Olney would not have anticipated. The code was a prototype for a rudimentary robot brain and would act as his template. But like most programmers, Olney had his own style and would sometimes find working with someone else's code frustrating. The Robot Teacher and the Philip K. Dick android would need coding with a whole new level of sophistication. Hammons's code had not been written with such a project in mind. As useful as it was, it was not enough. Olney decided he would have to start from scratch.

As with Hanson's previous visit, Mathews wanted to maximize the opportunity for publicity, both for Graesser's lab and for the FedEx Institute, where he now held an administrative position. He contacted a journalist he knew at the *Commercial Appeal*, Memphis's biggest newspaper, and sent out a press release to all of the local TV and radio stations. The main event, a series of presentations by various speakers, including Hanson, was to be held in a large conference facility adjoining the FedEx Institute known as the Zone.

The Zone was a four-hundred-seat amphitheater tucked between the institute and the Department of Management. Paid for entirely by AutoZone, a Memphis-based corporation, it had all the functionality of the main United Nations conference facility in New York and was the largest venue of its kind in the world outside the U.N. From the outside, with its sheer, curved metal walls, it looked like some kind of high-tech silo. Inside it was reminiscent of the Galactic Senate building in *Star Wars*, where representatives of various planets meet to discuss the fate of the universe.

Each of the chairs was fitted with a console that included a microphone, voting buttons, a personal LCD screen, and slots for ID cards that could be inserted by attendees when they took their seats. Cameras operated constantly. An image of the speaker at the front of the auditorium was propelled onto a screen above the podium by larger-than-life cinema-quality projectors; the proceedings could be broadcast throughout the complex. To raise a question, an audience member pressed a button. This caused cameras mounted in the ceiling to swing into focus on that person's seat, beaming his or her face onto the massive screen at the front.

Hanson was scheduled to appear as part of a daylong event with a lineup of experts all discussing advances in technology and the interface between technology and society. The lights went out and the audience was thrown into darkness. Hanson stood at the lectern under the spotlights and auto-tracking cameras.

This time, Eva was fully operational. Hanson showed the audience how she smiled. It was a warm smile, and although he could barely see the people surrounding him, he knew that they were watching her; they were being charmed by her authenticity and her beauty. He made her scowl, look worried, and open and close her mouth. He talked about the speed with which robotics, AI, and other technologies were progressing and merging, and described developments in cognitive psychology and consciousness. These developments were driving the birth of a new kind of science called "social robotics," the science of how humans and robots interact. He predicted that machines would soon become integrated into our lives in a way that previously only other humans could be. As the involvement of these technologies increased, there would be a need to design machines with which we could interact in a natural way, machines that could fulfill social roles as well as purely functional ones. For this to happen, we needed machines that looked like humans, in a way that was not just flesh-deep. And if we were to learn how to build such humanlike machines, we would also need a science to study how people interact with this new technology.

During question time, Stan Franklin, the codirector with Graesser of the IIS and the head of computer science at Memphis, voiced his skepticism about some of Hanson's claims. By training, Franklin was a mathematician. His latest book, *Artificial Minds*, had been released the previous year and had received a favorable review in the prestigious journal *Nature.* Franklin did not doubt the possibility of thinking machines or even conscious machines. In fact, he had recently claimed at a conference that a computer program that he had created was conscious.

Despite all this, Franklin was a Turing Test purist. He knew the test's limitations, but he accepted its basic premise: that the function of the machine is important, and that its structure and form are not. It simply didn't matter what a machine looked like. It was possible for an ugly, formless machine to be smart, and it was equally possible for a lifelike machine to possess no more intelligence than a storefront mannequin. Franklin was impressed by Eva—he agreed that the artistry was stunning—but in his view, Hanson's beautiful sculptures had no relevance to science. They did not advance the goal of building human machines in any important sense.

Franklin observed that Hanson's robot could not think. Hanson replied that there was more to being human than thinking.

"Is there any scientific value in building things that look like humans, rather than behaving like them?" Franklin asked.

"Yes, I think there are lots of reasons for doing this," Hanson replied. "When we interact with things in our environment we interact more naturally, and form more natural relationships, with things that look like us. How does that affect human-machine interaction? What does that do to the attitudes that people have with the technology around them? These are questions that interest me. And once we have something that looks human, how can we make it communicate with us? That's what people are used to. That's what they need from intelligent machines. There is power in that mode of interface."

Hanson went on to say that a program that did not have a

human form could not interact with people in the same way that a humanoid robot could. And since interaction was a prime requirement for intelligence, even in the Turing Test, in some ways programs could never be intelligent. Franklin disagreed. Computer programs could be intelligent, whether they were being run on a humanoid robot, the internal circuits of a submarine, or a human-resource software application. The form did not matter.

Art Graesser enjoyed the exchange. Indeed, he agreed with Franklin in some ways. But regardless of whether the Turing Test was right or wrong, useful or not, or whether Hanson's work could accurately be described as "science" or "art," the idea of building an intelligent robot excited him.

Hanson told me later that he was not surprised by Stan Franklin's skeptical attitude, which he says is quite prevalent in research circles. Rodney Brooks, the head of robotics at MIT, makes fun of the goal of building human minds and suggests in all seriousness that building robot cockroaches would be not only more realistic but more worthwhile. Many modern AI researchers have embraced a definition of intelligence that centers on algorithms, functions, and pattern recognition—that is, software, not physical appearance. The idea of making an android is considered quaint, a naïve dream for people who have watched too much *Star Trek* and read too many sci-fi paperbacks. Understanding and replicating many aspects of being human, such as falling in love, smiling at a joke, or reveling in the touch of skin to skin, are not considered interesting problems in computer science.

Hanson believed that art had more to offer science than scientists appreciated. In his view, inventing a new scientific theory was a creative act. Naturally, the theory has to be rational and able to withstand scrutiny, but some kind of insight, some flight of ideas, is also needed.

How did Alan Turing come up with the Turing Test? The process seems no different than that by which Philip K. Dick invented the Voigt-Kampff test. One is science and the other art, yet they both percolated from the mysterious cauldron of the human imagination. Albert Einstein invented the special theory of relativity while looking out the back window of a bus at the town clock receding into the distance, or so the story goes. But how did it actually happen? If buses and clock towers lead inexorably to special relativity, why didn't everyone on the bus arrive at the same insight, all rushing off after work to write a paper for a physics journal? The reason is that there is much more going on in such an act than joining logical dots. Einstein had a creative mind that could dance from one place to another, finding interesting paths to new solutions.

Hanson was right about the scientific process being in some ways creative. But to him it was more than that. He believed that solutions should come before the explanations of solutions, and that artists were better at intuitively finding solutions than scientists were. He said to me once, "Here's a different way of doing robotics, cognitive psychology, the whole package: use the artist as a black box problem solver." He was talking about an engineering black box, not the kind in planes that is retrieved after a crash. The term refers to a machine, perhaps an electrical box with wires and buttons, that can't be opened, so its internal structure is a mystery. Wires go in and wires come out, and you can plug it in and see how it works, but you can't see what's going on inside. The only way to understand the black box is by tinkering: switching everything on or toggling one switch or two switches at the same time and observing what the machine does. You can describe the functioning of the box, and you can even have a theory about how it is built, but you can never know for sure. Hanson thought that the artist could operate as some kind of black box problem solver, finding solutions and leaving the persnickety job of explaining those solutions to the scientist.

This view was shaped by his journey from the world of sculpture to the world of robotics. He did not have an engineering or science degree, and so he started without any preconceptions about what could be done and what was impossible. He did what he wanted to, and quickly found himself in uncharted territory.

When he built his first head he needed to connect motors under the skin to act as facial muscles. Luckily, sculptors are taught about the muscular structure of the body, and anatomy textbooks feature plenty of detail. But then he wanted to make those muscles contract in coordinated patterns to create various facial expressions. Now he needed something that explained how emotions mapped onto muscles, which then mapped onto expressions. His search proved disappointing.

The most promising lead came from the research of Paul Ekman. Ekman had developed a method, called the Facial Action Coding System, for breaking down any emotional expression into its component parts. Ekman had a list of muscle movements he called "action units" that could be combined to create almost any anatomically correct expression. An involuntary, genuine smile, known as a Duchenne smile, can be identified by the way certain muscles cause the corners of the mouth to turn up and other muscles cause the cheeks to rise and crow's-feet to form. A fake smile involves a different set of muscles, and, unlike with the genuine smile, crow's-feet don't form next to the eyes.

While FACS is of immense aid to researchers investigating human emotion, in itself it does not provide a comprehensive mapping from emotion to physical instantiation. The system was intriguing and helpful, but it was not the key to making Hanson's robots emote in a natural way. Besides, formulaic coding systems were not his thing. Hanson was at heart an artist and tinkerer, ill-suited to the dry, technical approach that Ekman's work offered.

Hanson gave up looking for scientific guidance and instead relied on common sense, artistic intuition, and trial and error. By physically installing motorized "muscles" and their controls himself, experimenting with positions and pulley lengths and various combinations of motors working with one another, he was able to create android heads with programmable emotions. Looking back, he felt that if scientists acted more like artists or if artists played the role of scientists, the sort of journey of discovery he'd experienced would be more common.

"The artist is trying to create a social aesthetic, a social intelligence. The artist doesn't care if it is in keeping with existing theories about brains or minds or evolution," he told me. "The goal is for the work of art to function as a social human. Then, once that is done, you can reverse engineer the solution."

It is an appealing notion, the idea that artists and scientists can work together to solve problems and make new discoveries. It's unlikely to catch on, though. Modern science, like the rest of society, is becoming increasingly specialized. It is divided and subdivided into a vast patchwork of enclaves, where experts in one thing toil away at the problems specific to their niche. People who can wander from enclave to enclave—"Renaissance men"—are rare. Hanson embodies a curious intermingling of art, aesthetics, and engineering. You can't mass-produce him.

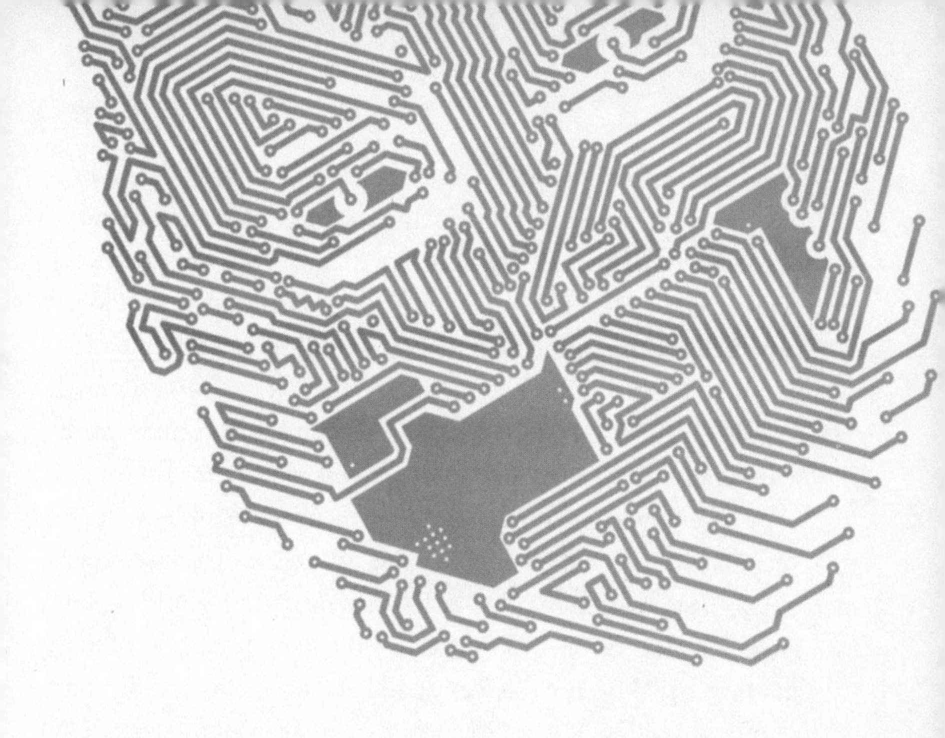

7. How to Build a Human Head

The grant money the team had applied for through the NIST Advanced Technology Program did not come through. They had proposed that if they had been awarded the grant they would build the Robot Teacher: Hanson would provide the hardware and Graesser's lab, the Institute for Intelligent Systems, the software. Funding bodies such as NIST like to hear about the potential practical applications of the research and what social need it will meet, so the team had explained that the Robot Teacher could eventually be used in situations where teachers were scarce. Funding bodies also want to know about the scientific value of a project, so the team had said they would study whether a robot teacher could be an effective school instructor. Nobody had looked into this issue before. For instance, should it look lifelike or should it be a robotic caricature? Would the fact that it looked and sounded human enhance learning or detract from it? Or make no difference? In other words, the team wanted to find out if realistic robotic teachers were worth the effort.

There were eight hundred other applications to the Advanced Technology Program. Of those, a panel of NIST reviewers selected fifty as a short list. The Robot Teacher made the short list. Of those fifty, the review committee conducted a final assessment and nominated twenty-five they considered to be outstanding and worthy of funding. Robot Teacher did not make that final cut.

Graesser, listed as the lead investigator, got an e-mail one morning from a NIST representative telling him the bad news. Rejection is normal in the grants game; Graesser had won grants before and had been rejected before, many times. At that moment he was an applicant or joint applicant on a dozen grant submissions with various researchers across the university and at other institutions. Judging from his experience in previous years, he expected about half to get through.

Hanson, however, took the loss hard. He was new to the game of research funding, and mentally he had already spent every dollar of his apportionment of the Robot Teacher project and figured out how to use the resources to bootstrap the Philip K. Dick android to it. Unlike Graesser, he did not have the backing of a university or several other large grants to fall back on. He was a graduate student who built robots in his apartment using his own cash. When he needed parts, he bought them from a nearby electronics store; when he ran out of money, he cannibalized parts from his old robots. Without income or resources from Robot Teacher, it would be hard for him to continue his work.

Chris Anderson from *Wired* had recently contacted Hanson as a follow-up to the New York dinner. Apparently the team that was organizing NextFest had decided to use the Philip K. Dick android as the centerpiece of its publicity campaign. But the android did not exist, and now there was no money to build it. The sensible thing would be to send a short, polite note to the NextFest people and let them know that he would not be exhibiting at the convention in Chicago next year after all. But Hanson had called in favors and used every connection he had to get the android into NextFest. He did not want to back out. He decided to hold off for the time being in case something else came through.

Hanson contacted Mathews to talk about the grant they didn't get and whether the people at Memphis were still interested in collaborating with him. Mathews assured him that they were. Hanson said that he was committed to going to NextFest and would do whatever it took to get there. Now that Robot Teacher was dead, their best strategy was to seek funding for the android project directly. It was not the sort of thing that typically got funded by major organizations like the National Science Foundation. But there are many ways to get a research project financed.

The problem was that it was not really research in the traditional sense—it was more development. The group wanted to apply and synthesize existing technologies in novel ways. Mathews soon came up with a possibility. The FedEx Institute itself had a research budget, and management had announced that they were currently taking applications for innovative research projects that could take place in the building.

Mathews had become increasingly involved in the FedEx Institute. He now held a paid position in the building's administrative section, with his own large office at the southern end of the third floor. He knew that the FedEx Institute was after cutting-edge technology and that its leaders wanted spectacular, splashy stuff that the local papers would be interested in. The people who made the decisions were in offices just around the corner from him; plus, Graesser loomed large as a major player on campus. There was no way they could lose. Sure, the grant would be only $30,000, small change for such an ambitious undertaking, but if they were frugal that should be enough to get the robot built and to Chicago.

Mathews mentioned it to Graesser, who liked the idea. With Olney and Hanson they put together an application for the FedEx grant, passing concepts and half-written paragraphs back and forth by e-mail. The process forced them to articulate more concrete plans for the robot than currently existed. Who would do what exactly? How would they build an android?

A plan took shape. The final document, sent in only hours before the deadline, merely hinted at this new level of specificity, at the map that the team members had created for the months between that moment and the opening of NextFest, in June.

It was a daring plan, far more ambitious than might have been evident from earlier discussions. The starting point had been simply to build a robot head that could talk, but they now thought they could do much more than that. For one thing, if it talked, it might as well be able to listen and to engage in natural conversation. This could be accomplished with real-time sound decoding and speech recognition. A microphone would function as the robot's "ear," taking input and sending it to a central processor, which would construct responses on the fly. Those responses would be passed to another program, one that converted text output from the dialogue manager to acoustic output like a human voice. This would then be passed to a speaker, and the robot would "talk." The words would simultaneously be converted into muscle actions, which could be expressed through movements around the mouth and elsewhere on the face via the activation of the motors under the robot's skin. The team would mount a camera that would function as the robot's "eyes," and connect it to software that could recognize faces and perhaps other nearby features, so that it could address people by name and comment on the immediate surroundings. The robot would also have the ability to track people visually, and what it saw would be fed into the musculature, the tiny mechanical positioning levers, so that the head could turn to face whoever it was talking to or follow people as they walked around the room. In addition, another module would control the robot's "emotional" state. This would generate various facial expressions, as it spoke or listened, appropriate to the situation.

In short, they planned to create a robot that looked and acted like a human being, complete with conversational capability, facial expressions, and humanlike movements. It would be a life-sized, lifelike, humanoid robot. In other words, an android.

They submitted the proposal for the FedEx grant in early February. Again, rejection. The winning entry was a proposal by a colleague, Lee McCauley, a professor in the computer science department. He planned to build what he called an intelligent kiosk, an AI program that would take the place of a receptionist in the foyer of the FedEx Institute. A face on a screen would smile at visitors, just like a human concierge. A camera would detect when someone approached, and a voice synthesizer would ask the visitor if he or she needed help. The visitor could interact with the kiosk either through a touch screen or through speech. To answer a query, the kiosk would access a database of employees, labs, rooms, and events. It would be able to give directions, responding to questions like "I have a meeting with a scientist here today. How do I find the meeting room I have to go to?" or "Can you tell me more about the Institute for Intelligent Systems?"

It was clear why the intelligent kiosk had won. McCauley's project had all the benefits the team had touted for the android project, plus the added advantage that it was practical—if it worked, it would be installed in the lobby, for every visitor to the institute to see. The selection committee probably realized that it had the potential for spin-off applications and even pickup by many businesses. There was little prospect of that happening for an animated robot simulacrum of a dead sci-fi writer.

Hanson was now in a panic. He knew that in Memphis, work had already begun on integrating components such as the speech recognition, face recognition, and behavioral output. Olney had started on the AI that would generate conversations. Between Memphis and *Wired* and the Chicago NextFest, many people were relying on him. But he had no money to do anything, even something as simple as pay for materials and machine parts to build the robot head.

Just as the situation seemed hopeless, another option emerged.

In 2003, Hanson had founded a company he'd called Human Emulation Robotics, in the hope of using the company as a vehicle for attracting investors. A group of venture capitalists in New Zealand expressed interest. They were prepared to finance his work at an unprecedented level, but they had a stipulation: they wanted him to move to New Zealand. This would make collaborating with Memphis very hard and pose difficulties for finishing his studies. New Zealand was also a long way from everything and everyone he knew. He declined the offer.

He soon, however, entered negotiations with another group of backers. Hanson has never revealed their identity, except to say that the people involved were venture capitalists and that they provided $50,000 to develop his business. The money was not specifically for the Philip K. Dick android, but over the next twelve months that's where most of it went.

Planning moved to the next stage. All the components would be built at the same time: the head, the body, the AI, the cameras and other hardware, the controllers for the hardware, and even the room in which the android would live.

Olney was in charge of putting together the software, and his biggest challenge was creating the AI that would hold conversations with people. This AI needed to be not only convincing but a convincing portrayal of the author himself. Olney had an idea about how to approach this. One way or another he would need to use Philip K. Dick's original works, as well as transcripts of interviews with him. Copyright for all of Dick's work was held by Dick's estate, which was made up of his three children, Laura, Isa, and Christopher. His two daughters managed the estate.

Mathews did the legwork, but Hanson became the main

contact with the trustees. He worked hard via mail and e-mail to reassure them that his intentions were benign. He sent them press clippings about his past robot projects and tried to explain his goals. The daughters were cautious. They wanted to know exactly how their father's work would be used, and in what context. They wanted to know whether it was a commercial venture, and who stood to gain.

Only so much can be achieved with long-distance communication. Hanson knew that he needed to interact with Laura and Isa in person. They agreed to meet him for lunch in Los Angeles, at the bistro of the Beverly Wilshire hotel.

Over lunch, the daughters quickly turned to business and bombarded Hanson with questions.

Hanson worked even harder. His biggest advantage was that he really was a huge fan of Philip K. Dick; he knew a lot about the author and believed in the project that had brought him to southern California to meet Dick's children. He told Laura and Isa about his teenage love for Dick's fiction, how important Dick's writing had been to him, even how Dick had shaped his personal philosophy. He told them how their father's stories about androids had led him to his own fascination with robotics and had moved him from working with sculptures to working with androids. He told them that he wanted to pay homage to one of the masters of science fiction in a way that seemed suitable and right: by building an android in his likeness.

Hanson also spoke to them about VALIS, a conception of God that Dick had described in his book of the same name. VALIS stands for Vast Active Living Intelligence System.

"I believe that artificial intelligence will lead eventually to a vast active living intelligent system," Hanson told the sisters. "I want to create friendly, benevolent intelligence that might become the next generation of humans: the *transhumans*." This, he said, was in accord with their father's vision. He hoped to contribute to a future world of superintelligent creatures that were somewhere between human and machine.

He sensed that the daughters were coming around to the idea. But still the questions came: What kind of robot would this be? Would it walk? Would it talk? If so, how would that be achieved?

"From a technology point of view," he told them, "this combination of speech recognition, face recognition, robotics, AI, and art is at the forefront of what can be done."

He talked about his trip to New York and how everyone at *Wired* was excited about the project; he explained that the magazine had already asked to feature the android in an exhibit at its upcoming NextFest expo in Chicago. They wanted to know if it was just him working on it. No, he assured them. He, Hanson, would be building the head, but there was a team of people involved, including Andrew Olney at the University of Memphis, who would be putting together the AI.

The AI was something Hanson wanted to address in his meeting with Dick's daughters, since Olney wanted to use the works of Philip K. Dick to construct a model of his mind and did not want to breach copyright. There was also the issue of the author's image. Hanson wanted to make a sculpture of the head of Philip K. Dick based on photographs of the man. His legal advice was that he should get permission first. He also felt it was right that the estate be consulted and asked for its approval, regardless of the legality.

Dick's daughters still had reservations and wanted a second

opinion from someone they could trust. They asked Hanson to talk to a friend of theirs, the Hollywood producer Tommy Pallotta, who, like Hanson, was a Texan. Back in Dallas, Hanson met Pallotta at a café, where again he explained what he wanted to do and why. Pallotta was persuaded. When I later spoke to Pallotta about the conversation, he told me, "David and I hit it off. The thing that won me over was that when we were talking about robotics and AI, we were talking about how it's the military-industrial complex that was funding most of the research and he told me he was resisting that, was trying to find a way around that, and that socially conscious robotics should be explored."

Pallotta told Laura and Isa that it was a worthwhile project. "If anybody should be made into a robot," he said, "it's Philip K. Dick. And it's probably what he would have wanted."

The daughters gave their consent to the project, allowing the team to use all of the works in Dick's estate and his likeness. They stipulated, however, that they wanted to see the final, complete work before it was shown to the public and that they would retain the right of a veto. In other words, even after the whole thing was finished, if they did not like the android, they could withdraw their approval at the last minute and prevent it from ever being seen.

Hanson agreed.

By the spring of 2005, work was in full swing. The team had only a few months to build a complete android in time for the Chicago NextFest, but everything was in place. They had permission from Dick's estate, members with clearly defined roles, and enough expertise to execute their plan. And they had money.

Admittedly, it was not a lot of money. It would only be enough to cover materials such as motors and plastics and expenses like transporting the android to the Chicago convention hall. Everyone on the team was working for free, devoting their time and energy for no other reason than the chance to build something extraordinary.

Hanson began spending most of his days in his apartment, which morphed from a rather messy bachelor pad into a rather messy robot laboratory. He had some materials and tools left over from Eva and other previous projects, but not enough to build a complete robot head. Plus, he wanted to do this one differently. He would use higher-quality materials, more finely calibrated motorized muscles. In the past, the technical expertise had come exclusively from Hammons's programming contributions. This time Olney and the others in Memphis were taking care of the software, but Hanson needed more help with the hardware, too. He had until now bought parts from the hardware store and installed them himself. But his being a solo operation had an impact on quality and durability. They were homemade robots, and they broke down a lot. This project called for a higher level of professionalism.

Hammons was connected to the Automation and Robotics Research Institute (ARRI) at the University of Texas at Arlington, so Hanson asked if the lab could help. There was already considerable interest at the lab in Hanson's project, and its staff was happy to get involved.

The involvement of the ARRI lab would make a big difference to the quality of the physical parts. This robot would not have $4 motors from the local hardware store; it would sport custom-built machine parts. And there was something else Hanson wanted to do differently. It was tedious starting each new head from scratch, going through the same motions every time. He wanted to streamline the process and start working toward mass production. The head of the Philip K. Dick android would represent the first steps in the emergence of a new production line for robots. He had heard about new technologies that might help, replication programs that could make as many copies of an object as you needed—so-called three-dimensional photocopying or digital modeling. He started searching around online and learned of a company in Maryland called Direct Dimensions. He called and was put through to Harry Abramson.

I asked Abramson about that first conversation. "It was apparent that David Hanson didn't have a clear idea of what he wanted us to do," he recalled. "He simply knew that, somehow, he wanted to store a digital copy of a skull."

Abramson had been with Direct Dimensions less than a year, and had been brought on to do "business development," to find promising new areas for the company to expand in. Hanson made it clear that he couldn't pay, but he proposed some kind of sponsorship arrangement. The project was ambitious, and Abramson thought there were kudos to be gained from it. It was also an interesting challenge, given the vague requirements and the need to try to find solutions where the client himself was not sure of what he wanted.

The skull itself would not be printed by Direct Dimensions; that would be done by another company, based in Dallas. Direct Dimensions would simply create the digital representation of the skull, using a sculpture that Hanson would send over as the guide.

It quickly became obvious that Direct Dimensions could not only help but solve additional problems for Hanson. The robot's skull would be invisible under the Frubber skin, but it played a crucial role, since it held and supported all the motors, wires, and pulleys that made the Frubber fold and stretch into human expressions. Hanson had attached these to Eva's skull himself, painstakingly fixing braces and brackets on the skull so that he could then hook on the motors and loop wires and other contraptions around and under the skin. But with a few keystrokes and clicks of a mouse, Direct Dimensions or another digital designer could simply add those braces and brackets to the computer representation. Then, when the representation was "printed" as a real, three-dimensional object, it would have brackets ready-made, already embedded into the skull. One more obstacle had been overcome, giving Hanson a shot of enthusiasm for the task ahead.

Abramson and his colleagues were excited about the android, so Hanson suggested that they come along to Chicago and help out at the NextFest display. The two discussed ideas. It occurred to Hanson that, in a sense, the people at Direct Dimensions were doing the same thing he was: they were both creating simulacrums of real-world things. He was making digital replicas of people, whereas they were making digital replicas of objects.

A few days earlier Mathews had suggested that the slogan for the Philip K. Dick display at NextFest should be "We Can Build You," after Dick's 1972 novel of that name. The novel tells the story of two salesmen of electronic organs and spinets who build simulacra of two historical figures: Abraham Lincoln and his secretary of war, Edwin Stanton. At one point the main character, Louis Rosen, attempts to convince the fake Stanton that he is an android, by pointing out details such as the fact that he was born around 1800, yet is alive in 1982. During that conversation,

Rosen reveals that he is beginning to suspect that he, too, is an android, that the real Louis Rosen is dead and he is a replacement built by his business partner.

Given the theme "We Can Build You," Hanson saw a chance to expand the display, to include the work of Direct Dimensions, as a way of giving the company extra exposure for its sponsorship of the android project. Direct Dimensions could "build you," too, in a sense, through its digital modeling programs. Hanson ran the idea past Abramson, suggesting that the company could have an area at the stand where they scanned people and turned them into computer simulations. Abramson liked it. He promised to have a team in Chicago in June.

With the logistics of the skull worked out, Hanson was ready to begin. But he faced an additional difficulty, one that had not been a problem for his previous robots: his subject was dead. Though Hanson had imagined meeting the great man in person someday, he had never had a chance to do so. He would have to base the work on photos.

He compiled a collection of pictures from online sources and from books, spanning Dick's adult life. There was one of him as a young man, cuddling a cat. In another he was leaning back on a couch, gazing at some distant point off to the left. A third shows him sitting on the ground, an unbuttoned shirt revealing a hairy chest. In a fourth he is standing in a green field, talking to a friend. In every one he looks sad.

Hanson decided to use the photo that appeared on the dust jacket of some of Dick's novels, a black-and-white portrait of the author in his early forties with an unkempt beard and a receding hairline, staring wistfully at the camera. It seemed to capture the essence of the man. It was also a well-known image of Dick and would conform to people's expectations of what a Philip K. Dick android should look like.

The first stage was to sculpt an old-fashioned clay portrait. He started by throwing a large lump of clay down on a board for the neck, then placed a larger lump on top and molded it into the rough shape of a head. From there he hollowed out the eye sockets and filled them with clay eyes, rolled lines of clay into the shape of ears and stuck them on the sides, added a nose and mouth. Once the basics were there, making it look like the man in the photograph was all about patience, an eye for detail, and skill.

When the clay sculpture was finished, Hanson admired his handiwork: it was a pretty good likeness. In fact, it was a fine work of art in its own right. All that remained was to turn that sculpture into a living robot, which necessitated the destruction of the sculpture before him. He had done this many times before. It always hurt.

He next encased the head in rubber. For this to work, the rubber had to be hot so that it was a runny liquid that could be poured over the clay sculpture. When the liquid cooled, the sculpture was covered in a thick coat of rubber. Hanson left it overnight to set. In the morning, it was time to extract the clay sculpture from inside. There was no way to pull it out from the rubber coating intact. The sculpture's only purpose was to create the rubber coating, so its job was now done. It had to be broken up and pulled out in bits.

Now Hanson had a rubbery mass that from the outside looked like a shapeless lump with a hole underneath. Its unseen, interior surface was a perfect inside-out version of the original sculpted face. In a sense, it was like a photograph's negative. It looked wrong, but everything important was captured in it.

Hanson used the rubber negative as a mold to make a second clay face inside. Because the rubber was a mold of the original clay sculpture, this new clay sculpture was exactly like the first one, which had been destroyed, but with one crucial difference: this time the clay was only an inch thick throughout the skull. The inside would be used to shape the skull; the outside would be used as a mold for the skin.

He made sure the clay was the right thickness all over the inside of the rubber mold, smoothing it with his fingers. Once it was finished, he turned it upside down and fixed it in place with a vise so that it could not move or tip over. Then he poured molten plastic down into the hole until it filled the rubber-and-clay sculpture. When the plastic cooled, it was hard. It was a smaller version of the original, and was a model of Dick's skull. Hanson called this the "positive."

Hanson peeled off the rubber, being careful not to damage it, and then removed the clay. Again, the clay had served its purpose and was no longer needed. For the second time, he destroyed a clay image of the author.

He now had the rubber exterior (the negative) and the plastic skull (the positive). He put the negative over the positive. There was a one-inch gap between them, the thickness of the second clay sculpture. Hanson held them upside down and poured a second plastic mix between the two to make a cast of the robot's skin. The mix was the liquid version of his patented plastic; when it cooled and hardened, it was Frubber. He pulled the rubber away and peeled the Frubber off the positive skull. In his hands was a replica of the skin of Philip K. Dick's face. Now he needed a skull to put it on. Certainly he had the plastic one he had made in the middle of the mold, but that would not suffice; it needed to be scanned and digitized. He took the skull and the face to the local FedEx, packed them into a box, and sent the box to Direct Dimensions in Owings Mills, Maryland. Then he waited.

When the package arrived at Direct Dimensions, Abramson passed it on to the engineers. They took the skull and the face to a workroom and laid them on a bench; there the contours were mapped with lasers and recorded by a nearby computer. Every point on the surface of the skull and on the face was now a data point on the computer's hard drive. The engineers used a common system for storing 3-D images known as IGES, or Initial Graphics Exchange Specification. IGES is the industry standard in computer-aided design (CAD), and it allows the data to easily be transferred from computer to computer, lab to lab, and software package to software package. The drawback is that the method takes a huge amount of memory. The engineers figured that Hanson didn't need or want so much data, so they compressed it into a smaller, lighter representation, in a format called NURBS (non-uniform rational b-spline). A couple of weeks after shipping the sculpture of Dick's face and skull to Maryland, Hanson got a CD in the mail with a NURBS representation of what he had sent.

He took it down to the ARRI lab. The lab had agreed to expand its assistance beyond producing parts to include working with the CAD representation from Direct Dimensions. Hanson found one of the programmers and handed over the CD. The programmer loaded the data into a CAD software program that allowed him and Hanson to play with the representation and modify it, adding holes and brackets to the skull. Hanson was watching weeks of labor be compressed into the space of minutes. This was so much easier than working with real objects. In the past, when he'd made a mistake he'd had to live with it. Here, the mistake could simply be undone.

Once Hanson and the ARRI programmers had redesigned the skull, the file was sent to the local 3-D printing company, which had a "rapid-prototyping machine" that could create any object from a CAD representation. Within days Hanson had the result: a new copy of the skull. It was exactly like the original but was made of harder plastic and included the lattice of holes and braces for affixing the facial motors. In theory, he could order a hundred identical Philip K. Dick skulls to be produced by the end of the week. But aside from the cost the skull was not a complete robot. It was merely the foundation upon which the robot head would be constructed. Hanson's dream of android mass production was still a long way off, but access to rapid-prototyping brought him one step closer. Perhaps one day he would have a production line of Philip K. Dick androids; for now, the challenge was building just one.

He held the Frubber skin that he had made from the original cast, turned it inside out, and attached small anchors to the inside. Then he inverted it again, lifted it above the skull, and slid it over. It was a loose fit, just as he'd intended. Now that the face was mounted, he needed to install motors under the skin so that the face could move. Without motors, it was still just a sculpture. Once motors were installed, it would be a robot.

Over the months that followed a routine developed. Working on some part of the android head, Hanson would realize that he needed a part—say, a particular kind of motor or lever. He would go to the local parts store, pick up some supplies, and build the part himself in his apartment, just as he had always done. But instead of installing it directly into the android, he would take that part down to the ARRI lab and explain what it was for. The engineer there would then use it as a prototype to build a better version. Several days later, someone from the lab would call to say his part was ready.

Hanson installed an array of motors, wires, levers, and other mechanical devices to create machinery that could pull the skin in various directions to indicate emotions. To operate the motors, sweeping cables attached each motor to the appropriate anchors.

There were now three layers to the head: the Frubber skin, the motors and cables, and the skull. Each motor and its cable operated on a single part of the face, pulling in a single direction. The tricky thing was getting them to pull in the right direction. To create a realistic simulation of facial muscle movement, two things had to happen. First, each motor had to pull along a vector in line with the surface of the face. It could not pull up, down, or in, because this would look unnatural and wrong. Second, the motors had to do what muscles might do in a human face. That meant they needed to be arranged in roughly the same positions as real human muscles and pull in roughly the same directions. After all the hardware was installed, Hanson had to tweak and adjust the position of each of the motors and cables so that they were just right. Calibrating the motors was a time-consuming and tedious job that took weeks to complete.

Now all that was left was to make it work. The face had a complex muscular structure, but which muscles would convey which emotions? By now Hanson had a system of sorts for rigging the motorized muscles under the skin, but that system was still entirely in his head and based on his experience of what worked. He would lay wires, attaching them to the Frubber, looping them through the brackets in the skull, then hooking them to

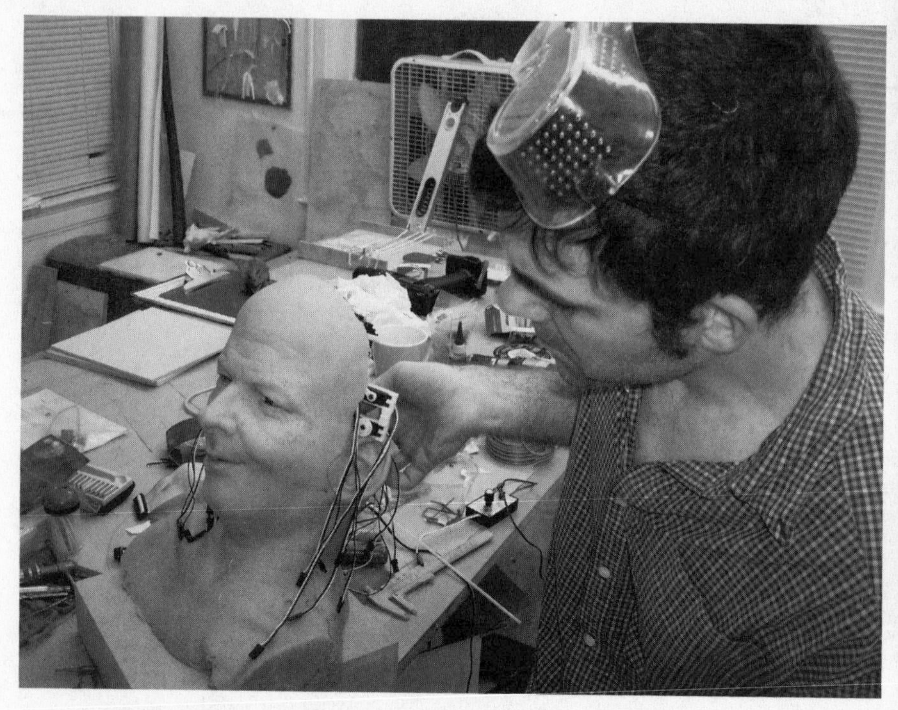

David Hanson working on the android's head

the motors. The motors, in turn, needed power cords to provide them with the "energy" required to perform their job as muscles, as well as transmission cords to signal to them what to do. The result was a tangled mess of leads, snaking in and out of holes and brackets and threading down, out of the robot's brain. Hanson had made skin for only the front half of the head. Face forward, it would look human, but from behind the machinery creating the façade could clearly be seen.

Once it was finished, he carried the whole thing into the ARRI lab. There, the engineers and programmers examined the placement of the motors and opened up the CAD representation once more. They modified the on-screen skull by overlaying it with a virtual Frubber skin, adding virtual motors underneath that matched the layout of the ones Hanson had installed. After some time, working as a team, three CAD programmers were able to create a simulation of the robot. With a keystroke or a mouse click, they could make the virtual robot purse its lips or raise an eyebrow; they could make it do anything Hanson might want the robot to do. The CAD simulator would be useful for testing out Olney's programming before putting it into the real thing. Hanson loaded the simulation onto his laptop and reclaimed his robot.

The robot head sat on the floor of Hanson's apartment, staring directly forward. Most of the motors were in position. The head needed a body, and to make one, Hanson again used his training from art school. He took plaster casts of the limbs and torso of a friend and bound them together with cloth, creating a mannequin. It was posable, meaning that it could be placed in realistic human postures, such as sitting in a chair with its hands in its lap or its legs crossed.

A package arrived from the daughters of Philip K. Dick. Inside were a few of the author's original possessions: several books, a snuff box, and some of his clothes. Hanson assembled the body and dressed it. The books and other curios were set aside for the time being. They would be used in the re-creation of Dick's living room, which was being built in Memphis.

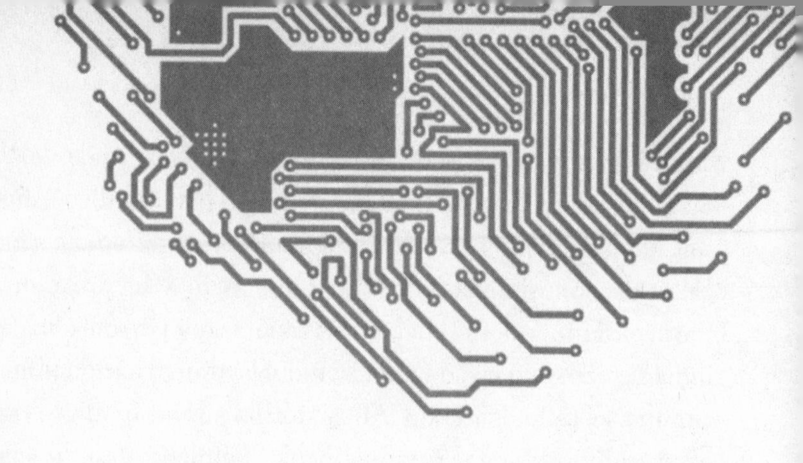

8. Life Inside a Laptop

That semester, in the spring of 2005, Olney taught a course in conjunction with Max Louwerse, whose specialty was linguistics in artificial intelligence and natural language processing. I asked Olney if I could sit in on the course. After checking with Louwerse, he told me that I could do the whole thing, including the assignments. He'd even mark my work, although as a postdoctoral fellow I was officially an observer, not a student, and so would not be getting a final grade.

Olney is a soft-spoken man who spends most of his time on

computer coding or reading books about computer coding. Standing at the front of a crowded room lecturing for an hour was not something he seemed naturally suited to do. He stood extremely still while he talked, as if directing all his attention to the task of speaking. Yet he had an economy of language and an intuitive ability to avoid detours through unnecessarily difficult terrain. He kept it interesting. To break the monotony of what was often a dry subject, he made obscure computer science jokes, which he punctuated with a wide grin. The challenge for the students, and for me sitting at the back of the room, was to implement the concepts we covered each week in the computer programming language Perl. Our first task was to re-create the very first conversational AI program, the brainchild of a man named Joseph Weizenbaum.

More than a decade after Turing proposed his Imitation Game, computers were still confined to NASA, military labs, and some of the more renowned research universities. At MIT, a professor with access to one such computer decided to have some fun with an innovative little program.

During the 1950s and '60s, new ideas in psychology and psychiatry were emerging. The rat-in-a-cage theories of behaviorists such as B. F. Skinner, who believed that studying thoughts and theories was unscientific because you could not see such things, were being superseded by cognitive psychology. There were also tectonic shifts in the world of clinical treatment. The ideas of Freud and his adherents were making way for a new wave of thinkers who called themselves humanists. One of the leading humanist thinkers was Carl Rogers, who had devised an innovative approach to counseling called person-centered therapy. The

MIT professor Joseph Weizenbaum thought he might be able to program a computer to act like a therapist in the style of Rogers.

Person-centered therapy adheres to the belief that it is better to help people find solutions to problems themselves rather than give them the answers. Rogers had written extensive guidelines on how best to make this come about and had explained the approach in several books. The best-known Rogerian technique is reflective listening, where the therapist merely reflects back what he or she has heard from the client. For instance, if the client describes a terrible day at work, rather than leaping in with suggestions about how to deal with the boss or consoling the patient, the therapist might say, "So you had a really bad day, huh?"

In traditional theories of therapy this does not add value, because it is really a restatement, but it does have several benefits. It shows that the therapist is actually listening to the client and taking note of what is being said; it gives feedback that the message was clearly conveyed; it allows the client to hear his or her own words in another person's voice and process them in a different way; and it prompts the client to further explore what was said.

Reflective listening can involve summarizing, focusing on emotional content, or simple reiteration. Weizenbaum thought he could at least get a computer to repeat back what was said, because this did not involve anything more than a basic knowledge of syntax: "I" must become "you," "You are" becomes "I am," and so on. So that if someone says, "I think you are really mad at me," this would be reflected back as "You think I am really mad at you."

Weizenbaum realized that a computer could be programmed to switch around the words in a sentence, and that in doing so it would be acting just like a Rogerian therapist. Such an act would not require intelligence or understanding or any of that other nonsense. All the computer needed were some rules, and perhaps some phrases that could be added at the beginning of sentences, such as "I am sorry to hear that" and, to spice things up, "What makes you think that?" Weizenbaum wrote a program that did just that.

Another key aspect of Rogerian therapy is that it focuses on emotions and relationships. Weizenbaum took care of that, too. He put in a range of keyword triggers for emotional content. If the program detected the word "mother" in the input, it would reply, "Tell me about your mother." Just like Carl Rogers or, for that matter, Sigmund Freud. If it detected the word "always," it would say, "Can you give me a specific example?" The rest of the sentence did not matter; nor did the answer. It was just a trick.

To round out the coding he also put in some simple question-asking features (asking is always easier than answering) and some random statements for situations where it was stuck. He called it Eliza.

Eliza became popular around campus but, to Weizenbaum's horror, people began to treat it as if it were a real, thinking being. His secretary would have conversations with this artificial therapist, closing the door so nobody could look over her shoulder and see what she was saying—or what was being said back. Here, from a paper Weizenbaum published, is an excerpt of a typical conversation between Eliza and a person.

GIRL: Men are all alike.

ELIZA: In what way?

GIRL: They're always bugging us about something or other.

ELIZA: Can you think of a specific example?

GIRL: Well, my boyfriend made me come here.

ELIZA: Your boyfriend made you come here?

GIRL: He says I'm depressed much of the time.

ELIZA: I am sorry to hear you are depressed.

GIRL: It's true. I am unhappy.

ELIZA: Do you think coming here will help you not to be unhappy?

GIRL: I need some help, that much seems certain.

ELIZA: What would it mean to you if you got some help?

GIRL: Perhaps I could learn to get along with my mother.

ELIZA: Tell me more about your family.

GIRL: My mother takes care of me.

ELIZA: Who else in your family takes care of you?

GIRL: My father.

ELIZA: Your father?

GIRL: You are like my father in some ways.

ELIZA: What resemblance do you see?

GIRL: You are not very aggressive but I think you don't want me to notice that.

ELIZA: What makes you think I am not very aggressive?

GIRL: You don't argue with me.

ELIZA: Why do you think I don't argue with you?

GIRL: You are afraid of me.

ELIZA: Does it please you to believe I am afraid of you?

GIRL: My father is afraid of everybody.

ELIZA: What else comes to your mind when you think of your father?

GIRL: Bullies.

ELIZA: Does that have anything to do with the fact that your boyfriend made you come here?

Even when you know the tricks that are being used, Eliza seems intelligent, and to be providing wise, thought-provoking responses to a confused girl. The conversation about her family is coherent, even intimate. Yet one of the participants in the conversation is not a person at all but a computer program that, whenever it heard the word "brother," was coded to respond, "Siblings can be trouble."

Still, there were some slipups. The girl says, "You are afraid of me." Eliza inverts this and tacks a stock phrase from its barrel of language fragments at the end. The result, "Does it please you to believe I am afraid of you?" doesn't logically follow, either on its own or in the context of the conversation. Yet the girl ignores this non sequitur and shifts the topic back to her own problems, convinced that Eliza is listening and caring.

There had been work in AI before Eliza. Turing and a colleague had dabbled in writing a program that played chess, and a team at Cornell was experimenting with early versions of neural networks. Eliza was unusual because it could talk. Eliza was not given a Turing Test, at least, not in the strict formulation described by Turing himself. But the test is surely meant to exemplify an idea rather than be the last word in assessing intelligence. The basic premise the test illustrates is that if a computer can fool someone into believing it is intelligent, then it is intelligent. This is what Turing was getting at. And according to this loose interpretation of the Turing Test, Eliza passed.

But Eliza works only if the person talking to it says things that can easily generate sensible answers and does not try to trip the program up. If you are out to prove how intelligent Eliza isn't, it can be done, as the philosopher Jack Copeland showed. In his book *Artificial Intelligence: A Philosophical Introduction*, he shares the following "conversation" he had with Eliza.

ELIZA: How do you do? Please state your problem.

COPELAND: I'm wondering whether computers can think.

ELIZA: Why do you mention computers?

COPELAND: Since you are one I thought I would ask you—can you think?

ELIZA: Did you think they might not be one you thought you would ask I can I think.

COPELAND: Let's talk just about you—not me. Can you think?

ELIZA: You like to think I—not you—don't you?

COPELAND: Pardon?

ELIZA: What does that suggest to you?

COPELAND: What does it suggest to you?

ELIZA: Tell me about what does it suggest to I.

COPELAND: That's nonsense—which of us do you mean by I?

ELIZA: Why do you think I mean by you?

COPELAND: The question meant, to which of us is the word referring?

ELIZA: Suppose the question meant to which of us were not the word referring.

COPELAND: I think you are an extremely stupid program.

ELIZA: Perhaps you would like to be an extremely stupid program.

Interactions like this have convinced computer scientists that Eliza was not intelligent at all but a computational sleight of hand.

This is only partly true. In a situation that is confusing or poorly understood, humans can also can say stupid things and make little sense. And just like Eliza, a lot of our verbal interaction is mechanical and programmed: "Hi, how are you?" "Fine, thanks, and you?" In such a conversation you are retrieving fixed word strings and fixed responses, just like Eliza.

Olney encouraged us to be creative with the Eliza clones we wrote for our assignment. After all, what's the fun in re-creating Eliza exactly as Weizenbaum had written, as an earnest therapist? The possibilities for new personalities are endless. You could have your clone recite poetry, for example. Alternatively, Eliza could be transformed into a rustic farmer, responding only to words involving weather or farm animals. Or a banker discussing Fortune 500 companies.

A famous Eliza variant is Parry, created in 1972 by Kenneth Colby at Stanford University. Parry, a paranoid personality, would say things like "I have incriminating evidence against the Mafia" and "They want to get false evidence to frame me." Conversations with Parry were fraught with suspicion and fear, revolving around conspiracies and organized crime.

Students created grumpy Elizas and movie-buff Elizas. I decided on a lighthearted approach and turned my program into

Eliza's sister Electra, a flirtatious female entity inside my laptop. It would make salacious comments: "Just talking to you is frying my circuits!" and "Wow, where did you get that sexy typing style?" and "Format my hard drive. You know you want to. And to hell with the consequences!" But Electra's personality was of secondary importance. The main thing was that the program worked, that it received inputs and gave responses.

At the next class, Olney talked to us briefly about our projects. They were good, he said, although many of them—mine included—did not clearly separate the data from the algorithms. For a small program such as an Eliza it did not matter, but as the program got larger the code would get messy and harder to keep organized. In other words, if it was not neatly organized it would not scale well.

Someone asked Olney if he thought the programs were intelligent. "That's a good question," he said and grinned, enjoying the confusion before him. Eventually the consensus among the students was that Eliza was not intelligent.

But what did it lack? What was the missing magical ingredient that would transform Eliza from a program to a self-aware, intelligent entity? Maybe a bigger version with more rules and keywords, a vast Eliza that had a response to every word and to thousands of phrases, would be intelligent.

Olney explained that there is a range of opinion. Some scientists have argued that it is simply a matter of scale, and that a big enough program will indeed blast through the intelligence barrier through sheer force of size. Others say that a set of rules will never do the job, no matter how large and complex, and that we need a completely different approach. Neural networks and biologically realistic models of the brain are a popular alternative with this group. Then there are those who maintain that machines simply can't be conscious and that we should give up trying to make them so.

Since Eliza, AI research has come far. AI programs converse with us over the phone ("If you are seeking financial assistance, press or say 1 now"), they play chess against us, they move navy personnel from one base to another, they control robots that traverse the surface of Mars. In guiding complex machinery and making abstract decisions, AI has become smarter than us. But it does not work like the human mind.

Computers have a hard time with perception and motor tasks, the sorts of things an average human is on the way to mastering by the age of four—simple things like noticing a half-full glass sitting on a tabletop or catching a ball. This is why it made the news when a robot team in Japan spent hundreds of millions of dollars building a robot that could walk up stairs. That was all it could do. From a scientific point of view it was a breakthrough, but for people watching the nightly news around the world, it seemed like a pointless thing to be spending money on.

One of the biggest obstacles is that researchers have underestimated just how much knowledge an ordinary person has stored in his or her brain. And it's not just knowing the way home, the names of animals and plants and people, or the life histories of characters on television shows. The really hard knowledge to pin down is implicit knowledge, the stuff that people don't even realize they know: being able to list the number of things that could be placed in a mailbox, for example, or how to play card games. But there are harder varieties of implicit knowledge, too: knowing which jokes are appropriate in which settings; understanding (even if you have never done it before) that if you throw a shoe at a cake, it will probably destroy the cake. Such tasks are very hard for computers to figure out, which is why programming solutions for them is known as the "knowledge problem." Computers can't give an opinion on the quality of a poem or the significance of a news item. They can be programmed to give an opinion, but they're faking it.

In 1990 a competition was launched by the entrepreneur Hugh Loebner and the Cambridge Center for Behavioral Studies. Anyone who could provide a computer program that could pass the Turing Test would win $100,000. As Jack Copeland wrote, "A program that is genuinely indistinguishable from human speakers would obviously require a vast knowledge store, containing more or less everything that the average human knows. In view of this alone it seems unlikely that there will be any hot contenders for the Turing Test for a good while yet."

The Loebner Prize has been contested every year since but, despite a strong field of contenders, year after year, to date nobody has won it.

Olney was spending his days teaching, marking assignments, attending meetings of the language and dialogue group, and working on his dissertation. In the evenings he would ride his bicycle home from campus, have a vegetarian dinner with his wife, Rachel, then sit at his computer and build the mind of Philip K. Dick.

It was a tough problem, and Olney considered various strategies. Of course there was always the possibility of just coding up an Eliza variant with Dickian responses. If asked about time travel, for instance, the AI could make some cryptic comment about parallel universes. But that would not do. It would be cheap and lame and probably not work very well, and would stand in stark contrast to Hanson's machinery, which would no doubt be elegant and complex. This android was supposed to be a scientific enterprise and needed to cover new ground. Quite apart from that, however, Olney wanted it to be excellent. This was his first chance—and possibly his last—to completely program an android. It was his chance to live a science-fiction novel.

Another way to go was the knowledge-engineering approach. This would basically involve hand-coding the inner workings of Dick's mind and would require extensive interviews with people who'd known the author while he was alive, asking them detailed questions about his interests, habits, and mannerisms. This information would then have to be coded into an intricate array of routines and possibilities. Olney did not want to go down that route. It would require a lot of work and, besides, knowledge engineering was a rather old-fashioned approach. There were better, faster ways to build an artificial mind, and these were precisely why the copyright issue had arisen.

A method that came to prominence in the 1980s was machine learning, in particular the use of neural networks. The basic idea is quite simple: you write a program that simulates some neurons communicating with one another, making sure that they have some inputs and outputs and some rules for learning. Real neurons are subtle things with complex electrical impulses coursing across them. The neurons in neural networks are simple, sometimes as simple as an on-off switch. If you give them the right training, however, neural networks can learn to do some very tricky things. For example, they can pick out a face from a blurry picture, the sound of a submerged mine on a submarine's sonar, or suspicious patterns of behavior in bank accounts—classic problems of pattern recognition.

But while neural networks are very good at doing some things, they are very poor at others. One of their most notable weaknesses is language. Debates have raged in linguistic circles about how much language can be learned by such pattern detection, though even its proponents do not claim that neural networks can handle anything approaching actual conversations. They can do word associations, such as figuring out that the word "cat" is very much like the word "dog" and not much like the word "jacket." They can also do tricks like converting verbs into past tense: turning "run" into "ran" and "watch" into "watched." That's about the limit, though. Conversing with a person is out of their league.

Olney had been involved with Graesser's lab, developing natural language systems, systems that could converse with humans in various ways. Systems like AutoTutor. These were usually hybrid creations, built with a mixture of theoretical approaches. Olney's attitude, and the prevailing approach in Graesser's lab, was that when trying to make real systems, too much theoretical purity only gets in the way. Once you have something that works, you can then add complexity and nuance and start researching interesting questions.

Graesser's lab was heavily involved in research into a new way of getting computers to understand words. Latent semantic analysis, or LSA, had recently been developed in Colorado at a lab that had close ties with Graesser's lab. LSA is way of capturing the semantic content of a word by looking at the range of contexts the word tends to get used in. It is a way for machines to understand meaning.

The basic method behind LSA is to first put together a large collection of documents. Each document can be as small as a paragraph or as large as a book. A computer then runs through the documents, counting all the words and creating a huge grid, with every single word running down the vertical axis and every single document running across the horizontal axis. This matrix is then filled according to which words occur in which documents. Words that are similar to each other will have similar patterns of occurrence, so they will have similar profiles in the matrix. To do this, the collection of documents required—linguists call it a "corpus," since it is a "body" of documents—has to be huge. A human could not read that much written text in any practical amount of time. But this kind of tedious activity—reading and sorting and keeping track of things—is exactly what computers are good at and can do very quickly. Through a data-reduction technique called singular-value decomposition, the matrix is then reduced to a much smaller size while retaining most of the information. The result is a bit like powdered milk. It contains all the essential ingredients of the original (other than water) but is a fraction of the size.

LSA works because it gives computers a chance to learn about words through exposure to lots of real-world language in context, just as children learn them. LSA played a central role in AutoTutor. It was a way for AutoTutor to determine whether someone was saying something relevant and whether it was accurate. Olney had been the chief programmer in implementing LSA into AutoTutor, using the context of a physics tutorial. In the case of the android, all the contexts would be generated by a bright, quirky, science-fiction-writing Californian. It would be an unusual take on the English language, to say the least.

The first step was to actually build the corpus. Olney returned to his sci-fi collecting days, snapping up Philip K. Dick books on eBay. Pretty soon he had an impressive library, from *The Game Players of Titan* through to *Ubik*, *VALIS*, and many others.

Once he had the books, he needed to get them into electronic form. Hanson reported that he had found a company in Washington, D.C., that could scan books on a large scale. The downside was that the books themselves could not be returned. All the books Olney had collected, his wonderful new library of sci-fi classics, would need to be destroyed, the spines cut out and the pages fed through a scanner. Olney put the books into a cardboard box and mailed it to the company. A week later he received a box of CDs containing all the books as digital files. He uploaded the CDs to one of the lab's Unix servers so he could convert the files into an LSA semantic space and create a model of Dick's mind.

When he was buying books and looking for any material that related to Philip K. Dick, Olney made a discovery. It turned out that there were several books and many transcripts of interviews with the writer. These dated mostly from the later years of his life. Apparently Dick had a rather open-door policy for interviewers, allowing friends, journalists, biographers, and others to visit and engage him in long, rambling conversations while a recording device captured it all. This seemed to be at odds with his supposedly legendary paranoia, but that's how it was. The collection of transcripts was vast. There may be more dialogue in print of interviews with Philip K. Dick than of any other person, alive or dead.

This discovery gave Olney a whole new idea for building the AI. The interview transcripts were loaded into a separate data folder from the novels, which were being used for the semantic space. Instead of breaking the interviews down into individual words, Olney broke them down into units of speech. One unit would be one person speaking, the next unit would be the response that followed, and the following unit would be whatever was said after that. He put it all into a database of dialogue utterances.

There were obviously two kinds of utterances in the database: things Philip K. Dick said in conversation and things someone else said to him. Some of the units were very small, just one-word lines: "fuck" and "yeah" and "wow." Others would stretch for hundreds of words.

Olney's program was going to depend on that dialogue database. When someone spoke to the android, the program would retrieve an appropriate response. If he could get every statement uttered by Dick in a recorded conversation into a database on a computer, the android could retrieve responses to almost any situation. Normally this would be impossible. Computer scientists would laugh at an attempt to build a language program that was set up to deal with any possible input. Getting or creating such a huge database of realistic responses is, on its face, absurd. Yet here it was, on Olney's Unix server. With so much transcribed interview material, such an approach might just be feasible. If someone sat down and said, "Hi," the android could look through all the times that someone said "Hi" to Philip K. Dick and see what sorts of things he said in response. If someone said, "Do you like sushi?" the android could look through the database and if Dick had ever been asked that question, it could simply retrieve the response and play it back.

Nobody in a transcript had ever asked Dick if he liked sushi. And therein lies the knowledge problem again. No matter how vast the database, you cannot have a store of all possible conversational moves because it is impossible to store all the knowledge that Philip K. Dick had. Besides, human language is infinite, so there were an infinite number of questions that could be asked of him. And yet . . . the dialogue database was big. *Really* big.

LSA could help here, bridging the gap between what was in the database and what people said to the android. Perhaps, Olney thought, LSA could find things that were not an exact match to the input but were at least relevant. With some rules to splice together bits of dialogue from different parts of the database, he might be able to create unique responses to some types of input.

Olney decided on a multilayered approach. The first layer would process the input and try to find an exact match in the dialogue database. If there was a match, then the android could simply say whatever Philip K. Dick had said in response to the same question or statement. If the first layer did not turn up anything, the second layer would try to do the same thing using LSA. This meant that it could find similar inputs in the database; they did not have to be an exact match. A third layer broke the input into sections and tried to find fragments of speech in the database that seemed relevant. It would take the best fragments it could find and piece them together into a single utterance. Other layers were tasked to do the same thing, in different ways. There are many methods for cutting up a string of words and putting them together again, and Olney made sure he had many angles covered.

As a final flourish, he also put in an Eliza-style layer that responded to keywords and to a handful of questions he thought might get asked a lot. There was no point in making the AI churn through all that computing power to assemble a response to a greeting like "Hi" or a question like "What is your name?" Some things could be hand-coded, Eliza-style. If someone said to the android, "Would you like to be human?" Olney's rules directed the AI to respond, "No, I am content with my robot existence." If someone asked, "What are you?" it said, "I am Phil, a male Philip K. Dick android electronic brain, a robotic portrait of Philip K. Dick, a computer machine."

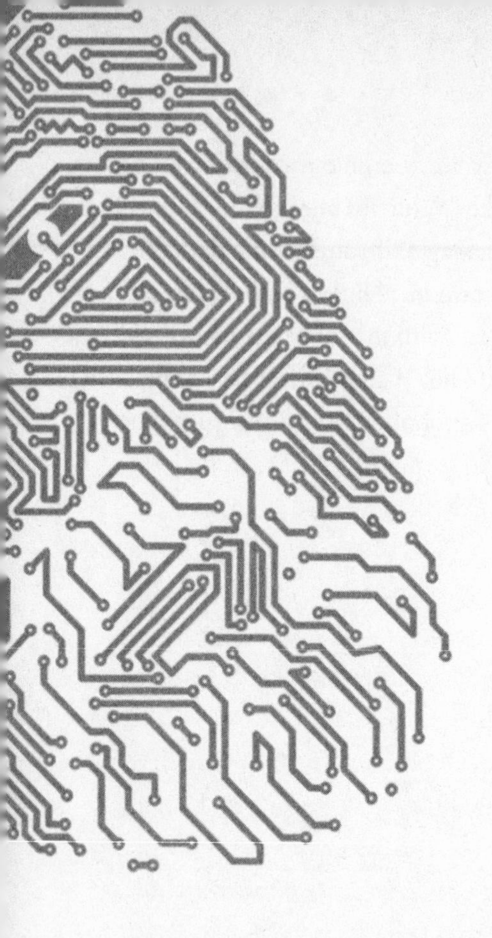

9. A California Bungalow, 1974

The best-known, most prestigious forum for presenting robotics research in the world is the annual conference of the American Association for Artificial Intelligence (AAAI). Hanson, Olney, Graesser, and others were cowriting a paper on the Philip K. Dick android for AAAI, passing drafts back and forth by e-mail. Halfway through the spring semester, Hanson learned that the paper had been accepted. He'd also been invited to enter the android into a robot competition that would be judged by luminaries in the robotics field. The conference would be held in Pittsburgh two weeks after NextFest.

The following week Hanson got a call from Tommy Pallotta, the film producer who had spoken to him earlier in the year on behalf of Philip K. Dick's daughters. Pallotta was the producer for an upcoming movie based on the Dick novel *A Scanner Darkly*, directed by his old friend Richard Linklater and starring Keanu Reeves, that would be financed by Warner Independent. Pallotta had told Linklater about the android, and the two of them were wondering if it could be incorporated into the publicity for the film. Hanson told Pallotta that the android had not been built yet. Pallotta asked if it would be operational by July.

"It's going to be at *Wired* NextFest in June, so it will be fully functional by then," Hanson said.

"Good," said Pallotta, "because Warner Independent has organized a major publicity event at Comic-Con in San Diego. Can you get the android there?"

This would be cutting it fine, as a week after NextFest Hanson would be taking the android to the AAAI conference in Pittsburgh. There'd be a mere twenty-four hours between the end of the AAAI meeting and the event at Comic-Con. He would have to go directly from Pittsburgh to San Diego, but it was doable, so Hanson promised that the android would be there.

Hanson was the only person on the project who had been to NextFest before, having exhibited K-Bot at the previous expo, in San Francisco. The experience had provided him with some valuable lessons. With Hammons's help they had developed K-Bot to respond to simple verbal commands and text input, but they had not anticipated the problems of running speech recognition in a public place. NextFest was loud. There were exhibits nearby, all generating noise, as well as crowds of passersby. The speech recognition did not work in that loud environment, meaning that K-Bot could be controlled only by text input. K-Bot did not have AI capabilities, so the failure of the speech recognition did not matter so much. This time, given the huge investment in AI and its central role in the new project, all that noise could be a disaster.

Through some contacts Hanson had made at the previous NextFest he had lined up some high-quality software. A company called Acapela was providing speech-synthesis software; this would give the android its voice. Another company, MultiModal Technologies, was donating its software to the android in return for public recognition of its contribution. It was an industry leader, and its software was far more accurate in transforming real-time speech into electronic text than what Hanson and Hammons had used the year before with K-Bot. Even so, it remained vulnerable to background noise. The people at MultiModal warned that, although the software was robust, very noisy input could degrade the results.

Hanson decided that a soundproof room might be the answer. Perhaps they could build a small, enclosed space, transport it to the Chicago NextFest, and display the android inside it. He told Mathews about the noise problem and his idea for solving it and asked if he knew anyone who would be able to build such a room. Mathews didn't, but he made some inquiries around the theater department at the University of Memphis. He learned that Mike O'Nele was the person for all-round set construction and design.

O'Nele liked to work on projects over the summer break, so he might be willing to do it, too.

Meanwhile, Hanson asked a friend of his, David Navalinsky, for advice on how he might go about building such a room. Navalinsky was employed to do set design and construction for campus drama productions at their university in Dallas, and Hanson was hoping he might offer his services. Navalinsky declined, but he recommended a colleague he thought might be ideal. There was a drawback, though, he said: the person lived in Memphis. Hanson told him that this was a lucky coincidence, as the other people on the project were based at the University of Memphis. Navalinsky and Mathews both made contact with O'Nele on the same day, within a couple of hours of each other. O'Nele told them both that he would love to be involved.

Hanson enlisted a friend to help him draw up a floor plan, depicting how he would like the room to be laid out. The room was to be small and rectangular, with a chair for the android, a couch for visitors, and enough space for all the apparatus required, such as computers and cables. He sent the sketchy blueprint to Mathews, who passed it on to O'Nele. Mathews asked O'Nele for a cost estimate. After some preliminary calculations, O'Nele gave Mathews a ballpark figure of $4,000 for the materials needed. Like everyone else on the project he donated his time.

The blueprint was O'Nele's starting point. For the next two months, Mathews was O'Nele's main contact, relaying requests and questions about problems back and forth and managing the money for the materials. O'Nele suggested that the room should have windows, and after consultation with Mathews, Hanson, and Olney, the change in the plans was agreed on. Originally, the team had envisioned a windowless container, to maximize the soundproofing, but they reasoned that having such a setup might be too intimidating for visitors. With windows, people would be able to peek in to see what was going on, which might tempt them to come inside for a closer look. The windows would also add to the authenticity, making it look more like a living room and less like a box.

As he started making more detailed preparations, O'Nele realized that the cost would be much higher than he had initially thought. For one thing, the room would have to be installed in Chicago, and even though it would be there for less than a week, it would have to comply with Chicago City Council building and fire regulations. Chicago fire regulations were strict and specific. Because electrical equipment was going to be used, the room would need to have professionally wired power sockets in the walls. A hole in the floor with extension cords snaking through it would not be acceptable. The regulations also stipulated that every surface in the room and every piece of furniture would have to be covered in flame retardant three times. The flame retardant recommended to O'Nele was a product called Rosco W46 that cost over $40 a gallon. When the room was painted, the paint would also have to be mixed with flame retardant. O'Nele explained the problems to Mathews, who relayed them to Hanson, who agreed to cover the increased expenses.

O'Nele's father often visited Memphis during the summer break, when O'Nele was usually working on one project or another, preparing sets for upcoming productions in the university's theater. The elder O'Nele was a skilled craftsman and often helped out. In the late spring of 2005 the two of them spent many hours in the workshop, building a room for the android.

Mathews would visit from time to time, checking on progress and keeping O'Nele in the loop. It seemed that he had unofficially become the project manager for the Philip K. Dick android. O'Nele explained to him that a slight modification in the design—with slightly shorter walls—would save a lot of time and money. Hanson approved.

The structure's frame was made of timber. For the walls, the only material that complied with fire regulations that was of any practical use was Sheetrock. O'Nele wanted to use soundproof Sheetrock, but at over $80 a sheet it was prohibitively expensive. He opted for standard Sheetrock instead. Then he and his father filled the cavities with fiberglass, lining all the interior surfaces with a material called Soundstop. They applied another soundproofing material, called elastomeric stucco, to the walls.

Now that the basic structure was in place, they needed to make it look something like Philip K. Dick's living room. Since joining the project O'Nele had started reading up on the author. He'd learned about his science fiction and his paranoia. He'd also learned that Dick had lived in a bungalow in southern California during the 1970s. This was when Dick wrote *A Scanner Darkly* and when he began to develop some idiosyncratic beliefs about reality. It was while living there, in 1974, that Dick had the revelatory experiences that led him to write *Radio Free Albemuth* and *VALIS*. This was also the year when he realized the world we see all around us is an illusion that imprisons us: the Black Iron Prison. A stream of students, drifters, and drug addicts passed through his home at this time, and to them Dick became something of a guru.

O'Nele decided to design the room as a California bungalow of the period. There was scant information about what Dick's living room looked like in 1974, or what he kept in it. O'Nele may not have had photographs of Dick's actual residence, but he could make an educated guess. The intention was to transport visitors back to 1974, where they would meet a replica of a man who believed that everything around him was an illusion.

The university library had a range of magazines with helpful pictures of the architecture of that era and directions on how to re-create it. One home in particular seemed to fit the bill. The article's title declared it to be "A California Bungalow, 1974." The interior color scheme was beige with gray wooden molding. O'Nele had seen that exact color scheme in pictures of several other homes of the period, so this was clearly a typical example

The android in Club VALIS

of the bungalow look he was after. Indeed, the look was so common that it seemed '70s California had virtually drowned in a sea of beige paint. The photos in the article showed track lighting and shag carpet and a sliding glass door that opened onto a patio. O'Nele considered installing the sliding door, perhaps with a small landing next to the room complete with fake grass, but the cost alone ruled it out. No matter. It was not essential. There were other ways of giving the room authenticity. Track lighting was installed in the ceiling and shag carpet laid across the floor.

O'Nele and his dad went to garage sales looking for period furniture. They found a '70s-style couch and got it transported to the workshop; it would be their centerpiece. They found some lamps in the theater department's props collection that appeared to be authentic, too. It was unlikely that the room they had built and furnished bore much resemblance to Dick's home, but it was a good representation of California living in the '70s, a time and a place with which Dick would have been comfortable.

O'Nele and his father installed wheels beneath the structure. When construction was finished, the bungalow was wheeled from the drama workshop and out onto the main stage.

As the project moved forward, students in the lab began to acquire and trade Philip K. Dick's books: *Do Androids Dream of Electric Sheep?*, of course, but also *The Transmigration of Timothy Archer* and a few others. The strangest of the lot was *VALIS*, which had been published in 1981, the year before Dick's death.

The novel *VALIS*—if you can call it a novel at all—is an extraordinary piece of work. It opens by introducing the main character, a hapless fellow named Horselover Fat, who narrates the story in the first person. Horselover has a disturbed and depressed friend, Gloria Knudson, who takes her own life despite Horselover's attempts to help her. The narrator is clearly a surrogate for the true author of the novel, Philip K. Dick: "Philip" is Greek for "lover of horses" and "Dick" translates from German as "fat" (it means, literally, "thick"). This in itself is not so unusual; it is common for fictional characters to be surrogates for the author, especially if the story draws heavily on personal experience. But then, on page 3, comes the following sentence: "I am Horselover Fat, and I am writing this in the third person to gain some much-needed perspective."

As a literary device it is audacious. Writers with an autobiographical impulse typically either are up front about their intentions or write thinly coded third-person stories that can best be described as semiautobiographical. But here the author starts with the pretense that the narrative is fiction, complete with false names to cover his tracks, and then confesses. On page 3. That is a disturbing revelation for a reader. It is the literary equivalent of getting into the passenger seat of a car, then realizing that the driver is on drugs. You don't know what's going to happen next, and you're wondering if you should get out while you still can. The impression is not completely unwarranted. *VALIS* is not an easy read. It is a meandering, dense work with lengthy philosophical and theological explorations. The dialogue between Horselover and his friends often exists simply as an excuse for Dick to dive into convoluted chains of reasoning about life, death, the universe, and sanity.

Dick had recently been coming around to the idea that his brain had been damaged by drugs and that he was not mentally well. He had accepted that many of his unusual beliefs and experiences were just the delusional products of a malfunctioning mind, and he was learning to dismiss them. But then there was the hernia.

In *VALIS*, Horselover has a revelatory experience. As he's sitting in his car in a parking lot, a pink laser beam shines down on him from the sky, and a machine intelligence, which he calls VALIS, contacts him. It tells him that his son is ill from a herniated intestine, which will kill him if not treated. Horselover takes his son to get urgent medical treatment, where the doctors discover that he does have a hernia and they fix it.

Several years before writing this passage Dick believed the real VALIS had communicated to him. While listening to the Beatles' song "Strawberry Fields Forever," he heard the lyrics morph into an urgent message: "Your son has an undiagnosed right inguinal hernia. The hydrocele has burst, and it has descended into the scrotal sac. He requires immediate attention, or will soon die."

Dick insisted that Christopher get medical attention right away, so he and his wife took the little boy to the hospital. The doctors discovered that Christopher did in fact have an inguinal hernia that required an operation. The intelligent voice from outer space had saved Christopher's life. If Dick really was just crazy, how had he known about the hernia? The event could not be explained away. It suggested to Dick that his religious experiences were not just hallucinations. They were real.

The verdict in academic circles these days seems to be that *VALIS* is a work of genius. It's not necessarily enjoyable, but it's expansive and ambitious, and that's the sort of thing that goes down well in literary studies. *VALIS* is certainly intriguing and unique. But is Dick acclaimed today as a significant literary figure because he was brilliant or because he was a spectacle?

What elevates Dick beyond being just another crazy writer with crazy ideas is his self-doubt. Dick was consumed by doubt: real, existential, solipsistic doubt, expressed over and over again in a thousand permutations. His characters are condemned to an array of situations of self-delusion: the android who thinks she is a human; the woman who wonders if her husband is an alien; the populace living underground, hiding from a war that is long finished; the paranoid schizophrenic, discovering concrete, irrefutable evidence that all his beliefs and suspicions are nothing but unfounded paranoia.

Even in *VALIS* Dick maintained what he referred to as the "minimum hypothesis," the possibility that everything he experienced was nothing but a hallucination. At one point he says of Horselover Fat, "Fat must have come up with more theories than there are stars in the universe. Every day he developed a new one, more cunning, more exciting and more fucked." These sound like the words of an author in control, not someone in thrall to his own fantasies. And yet those fantasies occupy a lot of pages in *VALIS*.

Hanson has no doubts about the value of the book, which was an inspiration to him as a teenager. "In *VALIS*, Dick explores the concept that AI is the next generation of humans," he told me, "that it will create a living mega-intelligence, a transcendental force that is capable of reinventing itself. And the smarter and more powerful it gets, it will reinvent itself in ways that we can't even imagine. Because there are two forces at work in the human race, the force of self-destruction and the force of creation."

Hanson was right that those forces coexist side by side in humans. In particular, the forces of creation and self-destruction coexisted in Dick himself, and these competing drives revealed themselves not just in his frenetic output but in the oppressive and dangerous worlds he invented.

VALIS is an exploration of both human possibility and human limitation. As Dick showed in his work and his life, we are capable of great things but our very human flaws and weaknesses can also be our undoing.

10. An Android in Memphis

The city of Memphis is jammed up against the Mississippi River on the eastern, Tennessee bank. When you approach from the west, you don't arrive gradually, as with most cities, with a slow, building crescendo of development. Instead it hits you all at once. One minute you're traveling across the lush Arkansas plain, dotted with silos and combine harvesters; then suddenly you're on the bridge, with office towers and high-rise apartments rushing at you.

Two long bridges—rather prosaically known to locals as the Old Bridge and the New Bridge—span the river, each over a mile long. A few miles into Arkansas, near a roadhouse and a cluster of chain restaurants, the highways from the two bridges merge, then split again. One, I-40, speeds over the horizon to Little Rock, then Oklahoma, and ultimately California, while the other, I-55, winds north along the river past small farming towns to Missouri, eventually leading to St. Louis and then Chicago.

On some January mornings, at the tail end of the brief southern winter, the Arkansas cornfields, usually green, will be covered with frost or snow. On those mornings, if the air is still enough, the surface of the Mississippi will reflect the city's buildings and the brilliant blue Tennessee sky, creating an image as idyllic as in any travel brochure. But the illusion is usually short-lived. Once the wind picks up, the waters stir and the river reverts to a bloated, brown channel of industry.

A week before NextFest, Hanson and his girlfriend, Amanda, packed their car, put a box with the android head on the back seat, and drove to Memphis. They crossed the Mississippi River in the sweltering heat of a June afternoon. Because of delays at the robotics lab with some last-minute parts, they were arriving two days later than scheduled.

There was still much to do. The various components of the android had not been put together; they did not know how long it would take or what problems would arise. Olney had written the software and connected the various applications that would run it without the benefit of the actual robot head that his software was going to operate. He would need time to link the android to the software and make sure that everything was working. Hanson's trip had been punctuated by calls from Mathews to check when they would get there.

To make matters worse, the head on the back seat was not finished. The motors that had been completed at the last minute had yet to be installed. There were cosmetic touches, too, that needed to be made. Hanson usually put makeup on his robots to reduce the tepid monotone of plain Frubber. He also wanted to give the android some facial hair, since Philip K. Dick was wearing a beard in the photos Hanson had consulted.

They drove directly to the university, where they met up with Mathews and Olney. Another man was with them. Mathews introduced him as Craig Grossman, the new director of the FedEx Institute. Grossman was more interested in the android project than his predecessor had been and had ensured full support from the institute for it. He was a science-fiction fan, although he preferred *Star Trek* to Philip K. Dick. He had once come to work on Halloween in costume, and had attended meetings and supervised the staff dressed as a crew member of the starship *Enterprise*.

The group crossed campus to the drama department and weaved past the theater's empty seats to the stage. There, on the stage, was a white building with a sloping roof and paneled windows. "This is it," O'Nele said.

They climbed onto the stage and O'Nele gave them a tour of the room, explaining where things were supposed to be placed and pointing out some minor changes from the original blueprint. Hanson looked around admiringly. This would be a good home for an android, he thought. He suggested that they name the room Club VALIS. Then he got the head out of the box and brought it into the room. The face looked like that of any ordinary middle-aged white man with pale, waxen skin. "This is Phil," he told them. "It's not completely finished."

That was certainly true. Philip K. Dick had been a hirsute man with a full beard and long, flowing hair. The head before them, on the other hand, was hairless. Not only did it not have Dick's trademark beard, it did not even have eyebrows. The face and the sides of the head were covered in something that looked like skin, but the back half of the head had no scalp at all, revealing the wires and motors within.

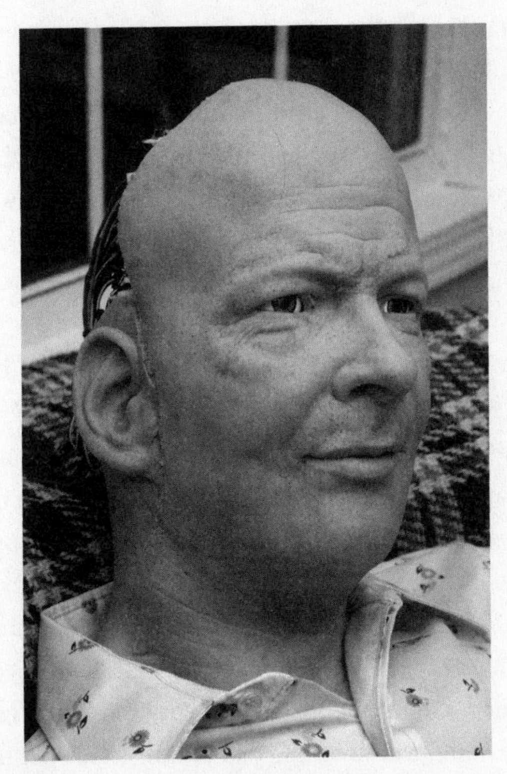

Android Phil before his beard was applied

Grossman asked Hanson about the back of the head. Was he going to cover it up with Frubber skin? Hanson said he would, but other things needed to be done first. He had to put on the beard, eyebrows, and makeup and finish installing the facial motors. Grossman invited Hanson and his girlfriend to stay with him and his wife in Germantown, a quiet outer suburb, east of the city. They had a large house with plenty of spare room, and there was a great Thai restaurant nearby. Evening was approaching and Hanson and Amanda had driven all day, so they left for dinner with Grossman.

NextFest was due to kick off in less than a week. Time was running out. Hanson and Olney started early the next morning inside Club VALIS in the empty theater, with coffee cups and muffins littered between laptops. They set up the android Phil. His head was mounted on a metal frame that was hidden from view by the body, which Hanson arranged into a human pose, sitting casually on the sofa. The room quickly became a tangle of cords and leads, computers, adapters, and screens. Power cables provided electricity to the motors that acted as Phil's muscles. A series of wires transmitted commands to the muscles from computers that were installed in the corner. Olney had brought several computers into the room. He needed a bank of three to run the basic android setup, which included several applications, operating in parallel, each with a heavy computational load. The facial recognition and tracking software alone required a dedicated machine.

Until now Olney had been flying blind, writing code for an android he had never seen. He had also installed the face recognition, speech synthesizer, and other modules and connected them to the dialogue software in a network of applications to run the android in real time. He had not once had the opportunity to test his programs to see if they would work on the real thing. It could all go wrong, and making mistakes entailed substantial risks. The software could command the motors to do impossible things or things that would break the machinery, such as telling two motors next to each other to pull the skin in opposite directions, causing the Frubber to tear.

The first step was to establish the acceptable range of movement for each motor. This was done by commanding a single motor at a time to move back and forth, until the limits of what that motor could do without damaging the robot were found. Those limits would then be set in the face-control software Olney had written. With this information they could create a map of the face and its mechanical muscles and then start the next step, which was figuring out how the motors could work in unison to produce emotions without ripping the face apart.

Eva, and Hanson's previous robots before her, had possessed a limited repertoire of expressions. They were realistic expressions that had been lovingly and painstakingly crafted through long hours of tinkering and experimentation, and they evoked feelings in those who viewed them. But without AI architecture underneath, a few expressions—smiling, frowning, looking downcast—were the limit of what was possible. The team wanted to make Phil capable of much more.

Like Eva, Phil's face would have a repertoire of emotions, but his was so much broader. He was capable of quizzically raising one eyebrow, for example, or looking surprised. And these expressions would not be performed for show but would be generated as a natural response to interactions with humans. At the same time, he would be moving his mouth in sync with the words he was saying. His utterances would be formulated in response to the environment and could not be predicted. While conversing, Phil would track whomever he was chatting with and orient his head to face them while speaking and listening.

These multiple processes would work in parallel, pushing and pulling the muscles into constantly shifting configurations unique to the moment and impossible to anticipate in advance. Just like a human's face, the android's face would be doing many things at once during a conversation. Phil could have virtually an infinite number of combinations of facial expressions and positions, making it difficult for the roboticists to ensure that various actions would not conflict with one another.

In four days the room and the android would be put on a truck and leave for Chicago. In those four days Olney had to test all the software and find and fix any bugs. He also had to get the entire hardware-software system of cameras, speakers, microphones, computers, motors, wires, and the various applications controlling them all working together.

Hanson was also still making last-minute modifications. Putting on the beard was a surprisingly finicky and painstaking task, and he had not even begun to cover the back of the head. Conscious of the time crunch and realizing that it would be enormously labor-intensive to install the hairpiece, Olney suggested that Hanson leave the rear of the skull exposed.

"It's fine without the hairpiece," he told Hanson.

Hanson stopped, sat back, and looked at the android. Olney was right. In a way, it was a statement about the whole project, the dual human-machine nature of the android, and the blurred boundaries between human and machine. It was a shocking

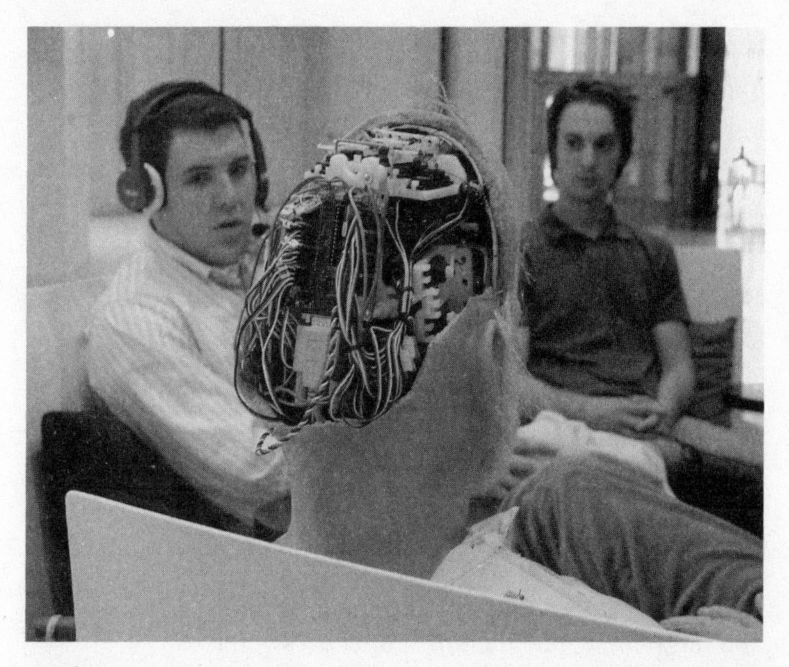

Graduate students Will Lancaster and Jonathan Nelson talk to the android at the FedEx Institute of Technology

visual effect. Phil looked like a head-injury victim from a road accident. It was almost as if someone had suspected that the robot was pretending to be a human and had ripped away the skin to expose its secret.

The time came to switch on Phil's power, all his modules and auxiliary devices, and see if he worked. They started the dialogue manager and waited.

"Hi there," said Olney.

"Hello," said Phil.

They ran some tests on the facial expressions; Phil performed perfectly. They tried a short conversation. Olney spoke into the microphone. "What is your name?" he asked, then waited for a response.

After a few seconds, Phil replied, "My name is Phil. What's yours?"

Olney was not surprised at the response. He had written "My name is Phil. What's yours?" into the code. The system was working.

Olney asked, "Where were you born?"

Phil said, "Phil was born in Chicago."

It was another programmed response. The real test would come with more spontaneous questions. Olney started making conversation, asking questions for which he knew there were no prewritten codes. Phil responded, and the two of them, Olney and the android, had their first, brief back-and-forth.

This is how the android worked: When someone sat on the couch next to Phil and said, "Hi, Phil," those words were detected by a microphone and transmitted down a wire to a computer, where they were sent to a speech recognition system. That system converted the words "Hi, Phil" from sound into text. That text string of letters and spaces, "Hi, Phil," would then travel along a cable to a second computer, where the dialogue manager operated. The dialogue manager had inherited from AutoTutor a processing structure called "hub-and-spoke" architecture. Like a bicycle wheel, there was a command center—the hub—that farmed out work to various submodules, which were the spokes. The hub of the dialogue manager first sent the text "Hi, Phil" to a module that converted it into an LSA representation, then on to the history module, which connected those words to everything that had already been said in the conversation. (Since "Hi" was the opening greeting, there would be no conversational history there, but the processor checked anyway.) Then hub sent the text string to a module that looked for instances of "Hi, Phil" in the database, then to another module, which looked for sentences of similar meaning to "Hi, Phil." The words "Hi" and "Phil" were also evaluated separately, as were hybrids formed by fusing a response to "Hi" and to "Phil" into a single long response. The dialogue manager kept track of all the possible responses it found and sent them through to a module that evaluated each candidate, scored and weighed it against the others, and selected one. This response was then sent to a text-to-speech converter. The vocal output was sent to a small speaker in the android's mouth. The speech was also converted into muscle movements, and commands were sent to the android to make the appropriate facial movements associated with human speech.

Phil responded, "Hi," and his mouth moved as a real mouth would to form the word. While all that was going on, the android was also processing visual input. The camera, Phil's "eye," was sending a stream of images to a dedicated computer that processed the surrounding visual field and scanned it for things that were likely to be a person's face. If it found one, that face was checked against a database of faces to see if it matched someone Phil "recognized," such as Olney, Hanson, or a friend of Philip K. Dick's whose photograph had been accessible and uploaded. If the computer recognized the face, Phil would address the person directly, for instance by saying "Hi, Andrew" or "Hi, David." The facial recognition program would also send the coordinates of the person who had entered the visual field to the robot controller, which would make Phil's head turn to face whomever he was talking to. If the person stood up and walked across the room, the head would turn to track him or her, watching the person walk. It took a lot of work just to get an android to say hello.

The next morning, Hanson and Olney switched on the lights in the living room and climbed up onto the stage. Inside Club VALIS, Phil was sitting on the couch in exactly the same pose as they had left him, staring into the distance. They powered everything up and started running tests. He malfunctioned just before lunch.

Hanson had installed heavier motors in the neck than throughout the face, since turning the head required more torque than wrinkling a cheek. The neck muscle motors were titanium alloy and had been custom-built by the ARRI workshop in Dallas. It was one of these special neck motors that caused the problem.

Hanson moved across the room and Phil, watching him, turned his head.

Olney asked Hanson, "What's that funny noise?"

They listened, but the funny noise stopped. It restarted a few minutes later, louder than the first time. Phil was talking to Hanson, who was sitting in front of him, when one side of Phil's face started vibrating. Then the entire face contorted into a horrific, impossible grimace. A loud buzzing sound and a faint burning smell registered as the skin pulled in bizarre configurations. Hanson turned around and shouted to Olney, "Kill the power!"

Hanson unplugged all the leads and connections to the head, removed it from the support frame, and inspected it. He peeled away the skin to look at the apparatus underneath and found that two motors had blown. The skin had been stressed, leaving stretch marks on the inside, but on the visible outer surface there was no obvious stretching or scarring. Some cables had been strained and needed to be replaced. Luckily, it seemed that there was no permanent damage to the hardware. All of this was repairable.

While Hanson replaced the destroyed parts, Olney tried to figure out what had gone wrong. A look through the code revealed no problems. He tried to re-create the exact moment that had caused Phil's paroxysm, but this was harder to do than he at first thought. Hanson and Phil had been engaged in free-flowing conversation, and because Phil kept track of the history of the conversation as he went along, it was not as simple as repeating the last thing Hanson had said. To re-create the moment, you needed to re-create the whole sequence of events that had occurred after Phil had been switched on that morning. Slight changes in wording caused the AI to go in different directions, to say different things and show different expressions. The state of the robot's mind was also influenced by the surrounding environment: the positions of people in the room who could be seen and tracked and the level of ambient noise, not just the conversational history.

A day and a half later, Olney announced that he had located and solved the problem. The face-tracking and head-turning module was not adjusting its movements to accommodate the module that controlled the facial expressions, and vice versa, leading to some instances where there was a conflict between the two. He edited some code, recompiled the modules, and reinstalled everything on the computers. In the meantime, Hanson had repaired the hardware. They plugged in Phil's head and powered up.

There was no self-destructing fit; he ran without mishap for the rest of the afternoon. But they had lost a day in fixing the malfunction and now had barely more than two days before they left for Chicago.

They mostly worked on their own, but there were sporadic visitors. Mathews dropped by at least a couple of times a day to report on the logistics for Chicago: how things were going with lining up a team of assistants for the display, the status of their accommodations, as well as issues regarding publicity. O'Nele checked in on the structure and, since he worked in the workshop behind the theater, sometimes just to see how things were progressing and have a chat. A couple of faculty members who knew about the project popped in, and Suresh Susarla, the computer science student who worked with Olney, came by.

Hanson would explain to visitors that Phil was a fully autonomous android. What he meant was that the android could interact with humans without being externally controlled. Phil would receive input from his camera eye and microphone ear and would generate a response based on that input. There was nobody sitting at a computer typing in commands, or crouched behind the sofa secretly controlling what Phil said or did. There was no puppetmaster. At least, that was the theory.

Complete automation was the team's goal for Phil, but the android had to be monitored closely in case something went wrong. Modules could crash, cords come loose, or the camera eye need realignment. The dialogue controller was easily disrupted by stray noises and would need to be reset. Hanson and Olney were finding problems and solving them day by day. The best way to detect bugs and other issues was to interact with the android as much as possible, but Phil's creators were often too busy fixing problems, changing code, and tinkering with hardware. They couldn't just sit there and talk.

They needed someone to do just that, however, so Mathews enlisted graduate students who worked in Graesser's lab to help out. Like acolytes in some kind of initiation rite, they were led down to the drama department, into the half-dark theater, and onto the stage, where they were led through a door cut out of the side of an odd white box. Inside the box were Hanson, Olney, and Phil, only two of whom were human. The students would sit and make small talk with Phil, and Phil would respond. Olney and Hanson remained off to the side, sometimes watching and listening, sometimes standing behind the android pulling wires, sometimes staring at a monitor. The two students who played this role the most were Suresh Susarla and Mathews's girlfriend, Sarah Petschonek.

Mathews had gone to Petschonek in her cubicle at the FedEx Institute and explained that the guys working on the android needed help. She'd protested that she knew nothing about

Eric Mathews and Sarah Petschonek pose with Phil

robotics, but he'd said that no expertise was necessary. Mathews was right. After she made her way from the bright, glass-encased institute to the dimly lit theater, Hanson and Olney asked her merely to walk in and out of the room. They wanted to make sure that Phil could turn his head appropriately to look at her at all times and that his tracker could follow her movements. She entered Club VALIS and Phil's neck swiveled, trying to face her.

"He needs to be facing the door head-on," Hanson said and repositioned Phil. "Okay, let's go again."

The next day, they needed someone to sit and talk to the robot so they could calibrate the microphone and the speech recognition. Petschonek sat there saying, "Hi, Phil" and "How are you doing today?" over and over. Nobody paid much attention to what Phil said in reply. Indeed, they did not pay attention to her, either. The roboticists were entirely focused on unseen processes on their monitors and the angles of moving motors.

One of Philip K. Dick's daughters, Isa, flew to Memphis that day to inspect the android. Hanson had promised the estate the final right of veto on the project, and now Isa was holding him to that promise. Isa met the key players in the lobby of the FedEx Institute—Olney, Hanson, Mathews, and Grossman. She was then taken to the room that housed Phil. When she entered, she witnessed before her a stunning likeness of her father, an android with wires and motors showing from the back of his skull. She was impressed by the resemblance, and by the amount of work that had clearly gone into creating the robot and the room. Hanson explained that the android could talk and, indeed, hold a conversation and invited her to take a seat. Isa tried to talk with Phil.

"It looked very much like my dad," she told a reporter years later. "When my name was mentioned it launched into a long rant about my mother and this one time that she took me and left him. It was not pleasant."

Olney explained that Phil had algorithms that drew on a vast database of Philip K. Dick's own words and created original responses on the fly out of pieces from that database. He told her that the things Phil had been saying were generated spontaneously by the AI programs running inside the computers on the floor and that nobody had told Phil to say those things. He and Hanson reassured her that this was not a joke at her expense.

Isa had reservations, but the android had moved her. It seemed to have captured something of the essence of her father. She gave her approval for the project to go ahead and be publicly launched at NextFest.

11. First, We Take Chicago

Everything was small and portable except for the soundproof room, Club VALIS. The android's head, its body, the frame that held up the head and body, the computers, the cameras, and the cables could all be thrown into the back of a large car or taken on a train, a plane, or even a bus if necessary. But the only way to move the android's bungalow was on the back of a truck.

With this in mind, O'Nele had been making arrangements to move the room to NextFest. A month earlier, before he had finished constructing it, he had called several trucking companies for quotes. He had been specific; he needed a stake-bed truck that could be loaded from the side and that could be covered in Department of Transportation–approved tarpaulins, and the driver needed to be hired to go from the University of Memphis to Chicago. Three companies said they had a suitable truck available for the dates in question and that they could take on the job. When he called back closer to the departure date, they all reneged on their earlier guarantees.

The first company he called back was the one he intended to use, and he believed he had a solid booking. But the man on the phone reported that they did not have any booking for him and no trucks were available for hire. O'Nele had noted the name of the person he had previously dealt with and asked to speak to him. The man on the phone came back with "He doesn't work here anymore. He didn't know what he was doing." The woman at the next company told O'Nele that they did not have any stake-bed trucks in their fleet. She did not remember speaking to him earlier and asked around the office. Nobody else remembered speaking to him, either. The third company did have a stake-bed truck, but all of their drivers were booked for those dates. O'Nele asked, "If I find my own driver, can I rent the truck anyway?" He was told yes, with the proviso that the driver must hold a commercial driver's license. It happened that O'Nele knew someone with a trucking license, his friend Ty Prewitt, who owned a local theater supplies shop. Prewitt agreed to drive the truck to Chicago and, when NextFest was over, drive it down to Dallas, where Hanson had organized a space to store Club VALIS on campus.

The day they were due to leave, O'Nele and Prewitt went

to the depot to collect the truck. The man on duty showed them the vehicle that had been set aside for them, a small, enclosed truck with a roller door at the rear. It was not a stake-bed truck. The man apologized and offered to cancel the booking, explaining that this was the only vehicle he had available that day. They decided to take it and make the most of the situation.

Prewitt drove the truck to the university and backed it down the winding campus alleyway to the rear entrance of the drama building. There was a loading bay next to the workshop where the room could be conveyed onto the truck. Hanson, Olney, Mathews, and Susarla were there to help. The android had been dismantled so it could be transported. The head was in a box out on the loading bay and the body was in another box next to it. Limbs poked out at odd angles, giving the impression of a mannequin hoping to escape its prison.

The original plan had been to drive the truck up next to the loading bay and lift Club VALIS across. That was impossible now. Prewitt backed the small truck as close to the edge of the loading bay as he could, jumped out, and opened the roller door. The layout of the alley prevented him from getting the truck any closer to the platform than a yard or so. Somehow they'd have to move Club VALIS across a one-yard gap and through the roller-door opening. The bungalow was wheeled through the workshop and out onto the bay. O'Nele measured the dimensions of the inside of the truck and the roller-door entrance and established that Club VALIS could squeeze inside. The structure, however, was taller than the opening in the door: the room had a beam sloping across its top, so that one end was too high to fit through the opening, though the other end would pass through easily.

Mathews had gone to Graesser's lab to find helpers. He soon assembled a contingent of around twenty graduate students, lab assistants, and assorted passersby who were willing to help. A grad student named Mike Rowe came up with an idea: slide Club VALIS over the gap, big end first, then drop it down and push it up at an angle, to wedge it past the roller-door entrance. Once the higher end was inside the truck, the room could be pivoted back to a level position and would simply slide in. Club VALIS weighed over a ton, however, so a lot of people would be required to hold it up from the ground while it was lowered, lifted, and pivoted.

A dozen people gathered in the gap between the loading platform and the truck. The room was wheeled to the edge, then pushed over. The group held it and let it drop a little, then pushed it up at an angle toward the back of the truck. The tip of the room cleared the roller door and slid inside. Rowe's plan had worked.

The group dispersed. Prewitt climbed back into the truck, drove it out of the alleyway, swung it onto the interstate, and headed north toward Chicago. Hanson and his girlfriend packed Phil's head and body into their car and left soon after.

Petschonek, Mathews's girlfriend, left in Mathews's SUV later that afternoon, toting Olney and a few others. Mathews had unrelated business in Memphis, so he had booked a flight and would meet up with them in Chicago. Petschonek drove midway into Missouri before stopping for the night at a cheap hotel off the interstate. Her group left at dawn the next day, arriving in midmorning at the Chicago Hilton, where they would be staying. Seven people—Olney, Mathews, Petschonek, Rowe, Susarla, and two other student volunteers, Will Lancaster and Jonathan Nelson—had traveled from Memphis to help with the NextFest display. As the only woman, Petschonek had the privilege of sleeping in a bed every night. The others rotated between the other bed, the couch, and makeshift accommodations on the floor.

After unloading their luggage into their hotel room they headed to Navy Pier, where NextFest was to be held and where they had arranged to rendezvous with Prewitt and the truck at ten A.M. Prewitt arrived at four in the afternoon, and Hanson arrived soon after.

This time, the group did not have to move Club VALIS themselves. NextFest had teams of staff members equipped with ropes, pulleys, and forklifts to help unload and set up displays. Once the room was out of the truck, a forklift drove it onto the convention floor and to the android's designated spot. The researchers trailed behind the forklift, lugging computers and boxes of gear. Hanson carried the head.

After the forklift had growled away, the team entered Club VALIS and began setting up the living room. They mounted the head and body onto the frame on the couch, arranged Dick's personal effects on a small table, lined up the computers in the corner, and plugged everything together with everything else. Hanson did some last-minute touch-ups to the robot, such as putting on the beard and makeup. Olney tried running some commands through the output box to the robot and discovered that some of the motors were not functioning properly; two were completely nonresponsive. Hanson had a supply of spare motors, which he quickly put to use.

Outside Club VALIS, the exhibition center was humming. Next to them was an exhibit from the Korea Advanced Institute of Science and Technology (KAIST) featuring a walking robot

Chicago NextFest 2005 with Club VALIS in the center

called Hubo. Walking is easy for humans but hard for robots. Simple robots that walk, like robot toys, don't have legs like people's legs. Without ankles and knees and hips, they lack the flexibility provided by those joints; they typically have two straight supports that allow a limited range of movement. Getting a machine with humanlike legs to be able to use them to walk turns out to be one of those surprisingly difficult engineering problems, like conversation and face recognition. We do it effortlessly, so it doesn't strike us as a tough challenge.

Hubo wasn't the first robot to walk like a person. That honor belongs to a long line of bipedal machines built by the Honda

Corporation starting in 1986. The first, named E0, was a pair of walking mechanical legs and nothing more. Later models E1, E2, and so on, up to E6, got progressively better. They had more sophisticated balance and were even able to walk up stairs. The series culminated in Asimo, a child-sized walking machine with a body coated in white plastic resin and an astronaut's helmet for a head. Asimo could walk; later versions could even run. Honda manufactured around one hundred units of Asimo, but it is no longer pursuing active research in the area. Development of newer, better bipedal machines has shifted to other labs, most notably at the government-founded KAIST, where Hubo was born.

Hubo was battery-powered, so there were no power cords trailing behind it. Likewise, the processing power was supplied by a nearby computer, with which it communicated wirelessly. Still, it could be tripped up. Twice during NextFest, Hubo fell over. Mike Rowe saw one of the incidents as he was waiting in line to view the robot. "It hit the floor hard!" he told me. "You've never seen people go into action as fast as those Korean scientists did when that thing landed. They were frantic. They were getting it back up in no time, reading their handhelds, and trying to figure out what went wrong." When prodded, the KAIST team said the fall was the result of a programming error.

Like Asimo, Hubo's entire reason for existence was to walk. It had been constructed to have legs with knee joints, a human-sized torso, and arms that were reminiscent of humans'. Despite the falls, Hubo was an engineering marvel. But Hubo had no head.

Hanson understood the sophistication of the machine, but he could not get past the missing head. It made the robot seem incomplete. Why make such a sleek, handcrafted emulation of a human and not put a head on top? This struck Hanson as a perfect opportunity; after all, heads were his specialty. Hubo was run by Junho Oh, whose team of graduate students was setting up the exhibit under his command. Hanson introduced himself to Professor Oh and invited him to visit Club VALIS sometime during NextFest, when he had a chance. Oh said he would.

On the other side of Club VALIS, a different kind of robot was being installed. It had two large orange metal arms, reminiscent of machines on a factory assembly line. Some people were lining up vinyl records near the robotic arms and wiring large speakers to an amplifier. Phil's handlers wandered over and asked about the display. They were told that this was Juke Bot, a creation of KUKA Robotics, one of the largest robot manufacturers in the world. It was a robotic DJ and would be playing music continuously for the duration of the NextFest expo.

Olney checked with Hanson. They had specifically requested a quiet part of the exhibition hall because of the fragility of the speech recognition package. Mathews found a NextFest official, explained the situation, and asked if their exhibit could be moved. The official said that he would find out. When he returned he reported that, unfortunately, the exhibits could not be moved this close to the conference's start. He apologized; nothing could be done. The KUKA Robotics team started testing Juke Bot. It was loud.

If they had been in the open exhibition hall, the noise coming from the Juke Bot would likely have rendered Phil's speech recognition useless. But because Phil was ensconced inside Club VALIS, there was still hope that he might function. If some noise leaked into the bungalow, it might degrade Phil's performance but not wreck things completely. They just had to hope for the best. The official opening was in an hour.

Down at the other end of Navy Pier there was finger food and wine and effusive speeches. This VIP opening was attended by *Wired* staff members, exhibitors, advertisers, media, public officials, and various folks who could loosely be described as famous. Neither Olney nor Hanson nor any of the other team members saw or heard or ate any of it. They were too busy finishing up their preparations in Club VALIS, troubleshooting and making last-minute fixes.

The guests drifted by in groups of two or three, carrying half-empty wineglasses, laughing. Phil was now operational, so some of the visitors conversed with him. The evening was a useful trial run. Club VALIS reduced the noise pollution from Juke Bot, but some of the relentless thump of dance music seeped in and interfered with the speech recognition. The roboticists disconnected the open-air microphone that was Phil's ear and connected a microphone headset.

The headset solved the problem of interference from Juke Bot's music, but it detracted somewhat from the experience, as now only one person could speak to Phil at a time. Despite the android no longer detecting all conversation in the room, the illusion that he was human, that Phil was a person, remained strong. This was helped by the shag carpet, the track lighting, the tattered brown-and-beige plaid sofa, the assortment of Philip K. Dick's books and possessions scattered over the coffee table; by the android's body, dressed in Dick's apparel, reclining like a drugged-out science-fiction writer chatting with friends; and, of course, by the face of the android itself, which even the author's own daughter had described as bearing an uncanny resemblance to the late, great Philip K. Dick.

Eventually the exhibition hall was quiet. Juke Bot had stopped pumping out music, the Korean guys had shut down Hubo, and the lights had been dimmed. The team returned to the Hilton—Hanson to the room with the people from Dallas, Olney to the Memphis room.

There were already people waiting at the front entrance when Phil's team turned up early the next morning. Inside, many of the exhibits were already operational. Hubo was standing at attention, waiting for its first scheduled demonstration of the complex activity known as walking.

Two men arrived. They introduced themselves as Harry and Mike from Direct Dimensions, the company that had scanned the skull. They had taken up Hanson's offer to be a part of the display, in keeping with the "We Can Build You" theme. Next to Club VALIS, they set up a computer on a small table, then pulled out a boxlike camera and mounted it on a tall stand. "It's a scanner," they explained. They would take three-dimensional pictures of people's faces and show them on the screen. Harry Abramson asked Hanson to step forward for a demonstration. He stood before the box. It snapped a picture and, before long, there was a 3-D image of Hanson's face rotating on the monitor. Hanson declared it to be a perfect complement to the exhibit. Some members of the Memphis team, not having been told in advance of this late addition, were unimpressed, but it was Hanson's call, not theirs. The android team entered Club VALIS, plugged everything in, switched everything on, and ran some quick tests. Phil was now powered up.

At the front of Navy Pier, the ticket gates opened and the crowds began pouring in. A tidal wave of the public washed through the alleys and aisles of the exhibition hall, past computer game manufacturers, futuristic vehicles, and laser shows. Some people made their way to Club VALIS. Although it was not close to the entrance, the android had been used in promotional material in the weeks leading up to the exhibition, and it was prominently featured in the NextFest program.

> Do androids dream of electric sheep? Now you can ask P. K. Dick himself. This bust relies on 36 servomotors to mimic the sci-fi legend's facial expressions, and features a polymer called Frubber that looks and moves like human skin. The bot uses motion-tracking machine vision to make eye contact with passersby, and best of all, artificial intelligence and speech software enable it to carry on complex conversations.

Within an hour there was a steady stream of visitors to Club VALIS: families with strollers, groups of teenagers, elderly couples, and lots of geeks. They were peering through the windows, staring in through the door, consulting their program guides, pointing, taking pictures. By noon there was a half-hour wait to speak to the Philip K. Dick android. At the start of the morning, the team had imposed a five-minute limit on conversations. By lunchtime they had shortened it to two minutes, but still there was a line of visitors standing beneath the *thump-thump-thump* of the Juke Bot, waiting for an audience with the android in the white box. It quickly became apparent that Phil was already the expo's most popular attraction.

Olney stayed by the computer bank while Hanson waved in visitors in like a host greeting party guests, inviting them to sit by Phil and put on the headset. The visitors would stare at Phil with a silly smile, glancing around at Hanson and Olney and the others in the room. Often a friend would also be there, taking photos or video or looking over Olney's shoulder. Mathews, Susarla, and the other students were busy keeping order in the growing line outside. They handed out brochures and explained the rules. ("You've got two minutes; try to speak simply and clearly.") They also ran errands and fetched fuel, mostly coffee.

The conversations with Phil varied from disjointed incoherent ramblings to moments of brilliance. Patterns began to emerge. Certain phrases and sentences were repeated by many people. In most cases, Olney had anticipated them and precoded some canned answers, some of which he had written in the third person. A common opener was "Hi, Phil," to which Phil said "Hi." A common question was "What is your name?" Phil would answer, "My name is Phil. What's yours?" Another was "Where were you born?" to which Phil would say, "Phil was born in Chicago." When asked what his favorite movie was, Phil would respond, "My favorite movie is *The Bicentennial Man.* Have you seen *The Bicentennial Man*?" When asked if he believed in aliens, Phil would of course reply, "Yes."

Many visitors were obviously hard-core science-fiction fans, some of whom knew an astonishing amount about Philip K. Dick. They asked questions about whether drugs influenced his writing, what he thought about Gnosticism, and whether Joe Chip, the main character of *Ubik,* was an author surrogate.

A man in black-rimmed glasses leaned forward, staring into Phil's eyes, and said, "Tell me about the meaning of the pink lasers." As he spoke, Juke Bot's volume suddenly increased as a song crescendoed. Olney, who was watching the speech-to-text conversion scroll down a monitor in the corner, saw that Phil had not understood a word. Phil sat in silence.

Olney told the man to speak louder because Juke Bot was interfering.

The man cleared his throat. "Tell me about your mystical religious experiences with the, uh, satellite from space and a pink laser beam."

PHIL: Military agents were behind it.

BLACK GLASSES MAN: Really? They shone a pink laser beam at you?

PHIL: Yeah, but it was different from all the other times.

BLACK GLASSES MAN: What other times? I thought it only happened once.

PHIL: I don't think you know what you are trying to ask me.

Black Glasses Man clearly wanted to continue to interrogate Phil, but Hanson had to politely end the conversation, as the man had run over his time limit and had to make room for the next visitor.

Another conversation, this time with the blogger Paul Jones, also known as "the Real Paul Jones," went like this:

JONES: What is your favorite color?

PHIL: Green.

JONES: What do you think about the president?

PHIL: Which president do you have in mind?

JONES: Where are we now?

PHIL *(looking around the room)*: We appear to be in my living room. *(Pause.)* It could be a simulacrum, though *(pause)*, but why would the authorities bother?

After recounting the event, Jones exclaimed on his website, "Now that is the real Dick!"

Phil made jokes sometimes, although whether these jokes were vestiges of actual jokes made by Philip K. Dick or unintentional was unclear. For example, a middle-aged woman arrived and, after seating herself next to him, nervously asked, "How old are you?"

Phil had a standard response to the question of age but for some reason diverged on this occasion, driven on a verbal detour by a fluke of the semirandom nature of his AI algorithms.

"I don't know," he replied. "How old are you?"

"I'm fifty-five," the woman said.

Phil's face tracker had locked on to her, and now his head rotated so that he was staring at her squarely.

"Oh no!" he said. "That's way too old for me."

The woman laughed. It was lucky she had a sense of humor. It would have been easy for her to take offense.

Some of the trickiest questions came from children. A boy wanted to know if Phil could sneeze. Another child asked Petschonek if the android would hurt him.

"No," she said, "it's a friendly robot."

On another occasion, she was inside Club VALIS when a small girl who had been talking to the android turned to her and asked, "Does it think?"

"No," Petschonek said.

"Then how can it answer my questions?"

Petschonek stared at the girl, tongue-tied, trying to think of an answer. Nothing came to mind.

By design, Club VALIS was a small, enclosed space, with the door and windows kept shut as much as possible to prevent noise from leaking in. By the afternoon it was, in the words of one of the students, "like an oven in there."

Thankfully, in the late afternoon the crowds thinned; then the front gates closed. Eventually the only people left in the center were the exhibitors, many of whom began exploring the other attractions around them. Beside Club VALIS, the Direct Dimensions contingent was still making instant 3-D scans of people. Several of the team members got themselves scanned. Hanson was first again. He stood upright for the camera; there was a click and Harry Abramson announced that he was done. Everyone crowded around the monitor on the table: there was Hanson's face in 3-D. Mike Raphael, the founder and CEO of the company, showed them how to rotate the image left, right, up, and down to any position at all, demonstrating that the head on the screen was truly three-dimensional and not some work-around. Next came Olney.

Olney, having not met or spoken with any of the people at Direct Dimensions before this day, had only a vague awareness of their role in building the android. After his scan, he peppered Raphael with questions about his business and area of expertise. While they talked, Abramson scanned Mathews, then Petschonek.

Next to them, the Hubo display had closed. The robot had finished demonstrating its prowess for the day. Professor Oh and his students came over to inspect Club VALIS. Inside the white box, Hanson and Oh talked at length about Phil. Oh wanted to know what sort of motors he was using, how many, and how they interfaced with the computer that controlled them. Hanson showed Oh the bundle of cords and wires that trailed behind Phil and across the floor, almost apologetically. Hanson knew that Hubo was not encumbered by wires, as it was powered internally and controlled wirelessly. But Oh waved his hand dismissively. Engineering involved trade-offs, and each machine had its strengths.

Oh asked about Phil's body. Was it part of the robot or just decoration? Hanson showed him the mannequin limbs. They had no motors.

"They create a sense of individual presence," Hanson said, "so in that way, they are part of the android because this is an attempt to simulate an individual, a complete individual, and they help the visitor re-create Philip K. Dick in his or her own mind." It was clear: Hanson built heads; Oh built bodies.

It had been a long day. That evening, once the crowds had dissolved, they all discovered that they were exhausted. They staggered back to the Hilton for some rest.

Mike Rowe was up early the next morning. He left to find breakfast and soon returned to the hotel room with coffee, a toasted sandwich, and a copy of the *Chicago Tribune*.

"Check this out," he said, opening the newspaper. There was a feature article on NextFest.

The article by Eric Gwinn was entitled "Future with Robotic Twist" and began with a gushing account of the spectacular array of new technologies on display at Navy Pier. Then this: "Drawing the most attention was a Philip K. Dick look-alike that quoted from the science-fiction author's books while fixing you with a gaze that moved seamlessly from placid to mildly annoyed."

The *Tribune* journalist stated that the android had been built by Hanson Robotics. It was an understandable error since the NextFest program itself described the android as being presented by Hanson Robotics, but this was not exactly true. Nobody in that room had actually spoken to Gwinn. It seemed he had met only Hanson and his marketing friend Steve Prilliman. The section of the article describing Phil ended with a quote from Prilliman: "As the population ages, these can be used as companions. In Japan, there's a shortage of young people for the elderly. This can provide the human touch."

The article focused on the android more than any other NextFest exhibit. It was a publicity coup. But Mathews was unimpressed. It didn't mention the Memphis team at all.

What was the point, he asked nobody in particular, of the university and everyone in this room contributing their time and expertise, and the FedEx Institute giving support, if only Hanson Robotics would be recognized for it? His jaw was tight as they walked to Navy Pier and started setting up for the day. Olney advised him to talk to Prilliman about it and then let it go.

When Hanson and Prilliman arrived, Mathews asked Prilliman to step aside for a private chat. Mathews showed him the article and asked if Prilliman had read it. He had. Mathews observed that neither the University of Memphis nor the FedEx Institute were mentioned; nor was Olney or anyone else from the Memphis team. Prilliman was taken aback. The interaction with Eric Gwinn had been brief and informal, and it had not occurred to either him or Hanson to give verbal credit to everyone involved. They had simply talked about the project. Prilliman had been focused on getting the journalist excited about the android in the hope that he would mention it in whatever he was writing. Clearly he had succeeded. He apologized for the omission and assured Mathews that future coverage would be balanced, giving credit to Memphis, too. Mathews later admitted to me that there was no ill will in Prilliman's omission.

The team had been busy the day before, but nothing had prepared them for Saturday. The *Chicago Tribune* piece had turned them from an intriguing entry in the NextFest program to the conference's major attraction. People who would not otherwise have attended NextFest arrived that morning with the sole purpose of visiting Phil the android.

A throng formed outside Club VALIS only minutes after the ticket gates opened. It soon arranged itself into a long line for the chance of an audience with Phil. Hanson shortened the time limit to one minute per visitor to cope with the increased demand. Even with this new limit, and Juke Bot thumping in everyone's ears, by lunchtime there was an estimated two-hour wait to speak with Phil.

The *Tribune* article had also caught the eye of other journalists. Rowe, after talking to people waiting in line, wandered back to the display to let Mathews know that there was a reporter from the Associated Press outside. Mathews spoke to her, making sure to mention Hanson Robotics as well as the University of Memphis. He had told Prilliman that all he wanted was balance, and he took the chance to prove it. In the afternoon, reporters from Reuters, CNN, a Russian news channel, and other media outlets stopped by.

The next day and over the following week, more than thirty stories appeared, plus countless blog posts. Phil was covered in newspapers as far-flung as the *Ottawa Sun*, the *Hindustan Times*, and the *Sydney Morning Herald*. According to the local *Chicagoist*:

> There really ain't much out there quite as eerie as an artificially intelligent Philip K. Dick robot looking you in the eye. And because it (referred to henceforth as "it" or PKDR for we shan't speak its evil name in full) spooked us so, we're sure it's evil and needs to be destroyed.

A reporter writing for the *Wall Street Journal* gave a mixed review, commenting that "the most advanced robot on exhibition was also, in my view, the most obnoxious." While he acknowledged the effort behind the project and the advanced AI on show, his interview with Phil had not been all he'd desired.

> Despite these enormous technological achievements, I was put off, perhaps because Mr. Dick—I mean the robot—not only didn't recognize me, but wouldn't answer my questions. . . .
>
> "Over the years it seems to me that by subtle but real degrees the world has come to resemble a PKD novel," the robot told me. "Several freaks have even accused me of bringing on the modern world by my novels. My writing deals with hallucinated worlds, intoxicating and deluding drugs, and psychosis. But my writing acts as an antidote, a detoxifying, not intoxicating, antidote." Hopelessly lost, I asked it what the hell all that means, but it ignored me and went on, staring at me all the while, following my increasingly agitated fidgeting with its incredibly

lifelike eyes, and employing the 60 sensors behind its face to sneer at me.

On television, Phil was featured on the *Today* show, the Discovery Channel, the cable channel G4, and the BBC. The online chatter was frenetic, particularly among science and technology bloggers. Robots.net reported that, at NextFest, "reality is stranger than fiction. Philip K. Dick has returned from the dead via FedEx. . . . David Hanson's skills provide the lifelike android head which reproduces Dick's facial expressions. Other experts have provided face-recognition and expression-recognition technology for the android."

NextFest closed late Sunday afternoon. As light from the setting sun angled in the windows, Phil was dismantled and packed into his black cases. The equipment, the computers, the cords, and the odds and ends were cleared out of Club VALIS. Their neighbor Juke Bot, silent at last, was wheeled away. Hanson made another visit to Professor Oh at the adjacent Hubo exhibit, this time with Eva, as well as Prilliman and some of the other team members.

Oh was keen to experiment with mounting a head on Hubo. Hanson brought out a small wooden box, attached it to Eva, and sat it near Hubo. The question was whether they would be able to secure the box to the top of Hubo's torso. Hanson and Prilliman measured Hubo, then the box. It seemed it might work.

They lifted Eva onto the torso, fastening the box with tape and some wire and concealing it behind a scarf. Eva's connection cords dangled down behind Hubo's body like entrails. Those entrails were plugged into a nearby power pack and a laptop that could send commands to Eva. The case with Phil's head tucked inside it served as a makeshift desk for the laptop, into which Olney punched some commands. Eva moved her eyes about, opened her mouth and closed it. Hubo had to remain stationary so as not to rip the cords out from under her.

Eva now had a body, and Hubo a head. It was not pretty. Hubo was unable to do what Hubo did best—walk—and Eva, precariously balanced on top, looked distinctly out of place. The torso was barrel-shaped and beefy, and even though Hubo technically had no gender, it came across as male. Eva, clearly a woman, was simply the wrong head.

Andrew Olney, David Hanson, and Professor Junho Oh, demonstrating the Eva android head combined with the Hubo walking robot

Hanson and Prilliman discussed how a head like Eva could be fixed more permanently to Hubo. Some kind of brace would be necessary, and since the robot was remote-controlled, the head would also need to be remote-controlled and would need to feed off Hubo's power source.

Despite the monstrous chimera before him, Oh liked the idea. There were only two lines of bipedal walking robots in existence, Hubo and Asimo, and neither had a functioning head or face. Certainly Asimo had a helmet mounted on top that looked rather like a head, but it had neither muscles nor facial features. The combination had not been done before, anywhere. Together, Oh and Hanson would create the first ever walking, talking robot.

Oh wanted to present something at a major exhibition in Korea later that year and asked if it would be possible to attach Eva to Hubo in the meantime. Hanson had a better idea: he would custom-build a new head for Hubo. The schedule was tight, but he was confident he could get something finished in time. Oh smiled. They would talk more.

In fact, Hanson had been thinking this problem over for the past few days and had already come up with the answer: the head could be in the image of Albert Einstein. He had seen a recent news item about the hundredth anniversary of the special theory of relativity, and it had struck him just how much had changed in a hundred years. Before now, his heads, like Eva, had been more in the mode of classic sculptures, but his success re-creating Philip K. Dick had made Hanson confident that he could attempt to re-create another famous dead person and pull it off.

A forklift came and carted Club VALIS to a waiting truck, destined for Dallas. The black cases containing Phil were loaded into the back of Mathews's SUV. Olney, Mathews, Petschonek, Rowe, and the others returned to Memphis with the android, while Hanson and his friends departed for Dallas.

Hanson was tired. The android display at NextFest had been a smash hit, more successful than he could have imagined. But on the flight home, all he could think about was this new idea, the Hubo project with the Koreans. The more he thought about the choice of Albert Einstein as the head of Hubo, the more excited he got. Einstein, the most iconic scientific thinker of the twentieth century, the inventor of the special theory of relativity and the father of modern physics, would be joined to a demonstration of some of the most advanced technology of the twenty-first century. A visionary genius from a century ago conjoined with a physical demonstration of the frontier of human accomplishment today. Now, there's a comment on the state of society and emerging technology, Hanson thought. That's art.

Amid the commotion of NextFest—the crowds, the fans, the reporters—in the corner of Club VALIS, Olney had been quietly wrestling with Phil's AI, trying to keep it running.

One of the unexpectedly tricky problems in building the android had been having a system that could respond in a timely, natural way. For humans, the rise and fall of vocal tones, the exhalation of breath, and a range of nonverbal cues help alert a listener that the roles have swapped and that it's his or her turn to talk. Like many things, a task so simple for humans is difficult for androids. Previously, Olney had built conversational systems where the user communicated by typing. When someone hit Enter the computer would start processing the input. But for natural, spoken conversation, the android had to figure out when the person had stopped talking and it was time to respond.

Olney solved this by having Phil's AI wait for a pause of several seconds in the speech stream before responding. One of the AI modules monitored the flow of speech to detect any pauses. If there was a long enough pause, this meant it was Phil's turn to talk. If there was no pause, or if there were only brief pauses in a stream of speech, then Phil would continue to listen. This was clever, but it also created a problem, because the dialogue manager's modules sometimes took several seconds to generate and output a response. In some cases, the person talking to Phil would say something else to fill the silence, so that when Phil finally did give his reply he would interrupt and talk about something that had been said up to half a minute earlier.

Olney's solution was to make Phil process the text while the person was still talking. The AI would begin assembling a response while waiting for the person to pause, incrementally processing the incoming language stream in small pieces as they arrived. When a pause did come, much of the computational work had already been done.

This system produced fast, relevant responses, but it also meant that the longer someone spoke to Phil without pausing, the more processing was involved. Longer inputs placed a strain on the AI as it searched through more and more fragments of text from the database and pieced them together; it also tended to produce lengthy results. At NextFest, even with the headset, the ambient noise from Juke Bot and random interjections from onlookers blended with the sounds the AI was processing, enough so that sometimes Phil would think the person in front of him was still speaking. He would continue to listen, continue to assemble ever more elaborate replies.

Unfortunately, these very long inputs were also causing Phil to freeze. After some investigation, Olney discovered that Phil was getting caught in a loop, adding more and more words and phrases to answers that were already impossibly long—sometimes thousands of words—and would become stuck. If left to his own devices, Phil would not start speaking again for years or, in the worst case, before the end of the universe. He would fall into an android coma.

A related problem was that a very large response would be formed, and Phil would start to talk anyway, continuing to add miscellaneous phrases to the ends of sentences even as he spoke, so that he could never actually finish what he was trying to say. In this scenario, if left to his own devices, Phil could speak forever. This didn't happen often, but the NextFest team came to refer to it as "one of his monologues" or "losing it."

Whether the problem was eternal silence or endless rambling, the solution was to empty the memory buffer. This was done by shutting down a process running on computer No. 2, then restarting the process while leaving everything else running. It was a dirty fix, but the situation was rare and Olney decided that he could come up with a better solution later, when the android was back in Memphis. In the meantime, while people were interacting with Phil, Olney was always lurking around the computers, ready to act if Phil started behaving strangely.

The NextFest program suggested that visitors ask Phil, "Do androids dream of electric sheep?"

It was an obvious question, and as Phil was actually an android, he seemed uniquely placed to answer it. What do androids dream

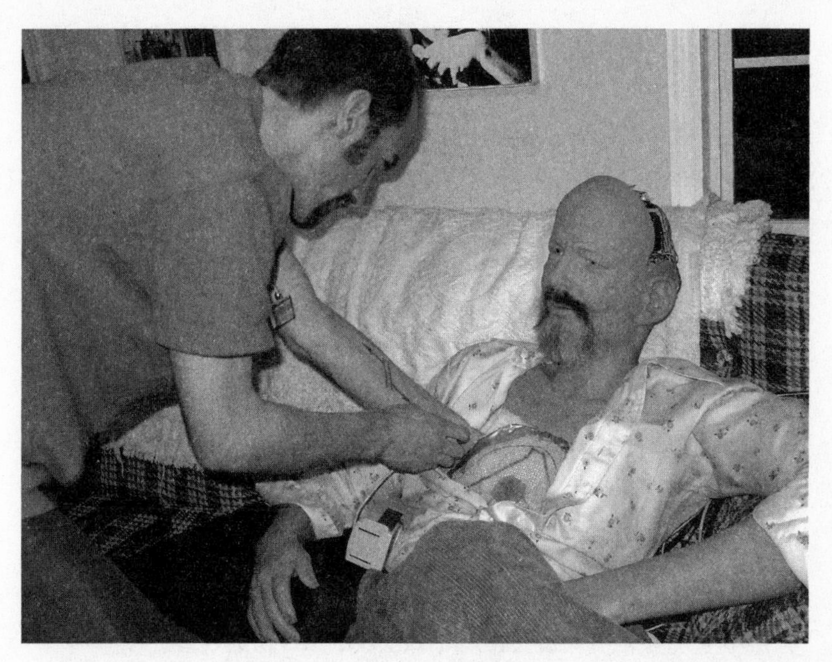

Andrew Olney making adjustments to the android

about, if they dream at all? To ask if someone or something dreams is an indirect way of asking if that being is conscious. One might as well say, "Tell me, do androids have souls?"

Though Olney and Hanson had written canned answers for many possible inputs, surprisingly, neither of them had anticipated this particular question. Since there was no programmed reply, Phil's AI modules had to search for the question in the response database. Even more surprisingly, it turned out that in the hundreds of pages of interviews and conversations with Philip K. Dick, there was no record of any person asking him this question. The database search for an exact match came up with zero hits. This meant that the dialogue manager had to fall back to the standard response generation for any unexpected or novel input: parsing, breaking down, and analyzing the input; finding numerous partial or possible responses; and piecing those together into something that seemed relevant.

The response Phil generated was not one of his short, witty ripostes. It was not short at all. The first time he answered the question, they feared he had gone into one of his rambling monologues. The second time, the answer was similar, though not exactly the same. Here is a typical answer from Phil to the question:

> Yeah, exactly. But I couldn't explain that feeling. I just couldn't give an explanation of it. But um, well, I remember one dream I had. In the dream Tess and I were in the kitchen on a high stool, we'd found a cereal box and on the back of the cereal box is extremely valuable information directed at us, and we were both reading it. And the relationship between that and *Ubik* is an obvious one. In *Ubik* Joe Chip is constantly getting personal messages from his dead boss Glen Runciter, for example on a matchbook cover. The ads were directed at Joe Chip, and that's precisely what I dreamed. And this kind of dream continued for months where written information like that was directed at me, although they weren't perhaps as conspicuously like *Ubik* as the cereal box one.

Perhaps some questioners might have expected him to talk about *Blade Runner* or the novel the film was based on, and on some occasions he did. But here he is talking about *Ubik*. *Ubik* is a novel that takes place in the near future, where corporations dominate and industrial espionage is rife. Psychics exist and are used as spies. Joe Chip is not psychic, but he is the leader of a team of psychics on a covert mission for his boss, Glen Runciter. The team is ambushed and Runciter seemingly dies in the explosion. But Chip receives messages from Runciter in bizarre ways, such as through graffiti on a toilet wall and on the side of that matchbook. In the interview that the android was accessing Dick had been trying to say that the inspiration for this aspect of *Ubik*, a dreamlike story in some ways, came from his own dreams.

It was not really an answer to the question, but the question itself is highly elliptical. The best insight we can get out of Phil is that androids may or may not dream of electric sheep, but Philip K. Dick certainly did.

12. Talking in Memphis

It was summer break and the University of Memphis campus was quiet. My footsteps echoed through the lobby of the FedEx Institute as I headed to the elevator. The bland, vacant face of the intelligent kiosk smiled and blinked from a plasma screen on the wall. Over to the left, by the eastern entrance, was a glass-walled enclosure, and sitting inside was Andrew Olney, hunched over a laptop on the couch next to the android. The door—a sliding glass panel—was open, and I let myself in.

"Are you going to switch it on?"

Olney replied without looking up.

"Yeah, there's going to be a big event tomorrow morning. It's being advertised on local radio stations today and maybe tonight's TV news, so there will probably be a crowd. The problem is, I'm trying to make some changes before that happens."

"Anything major?"

"Kind of. It's all to do with the speech recognition. You'd be surprised how hard it is to get a robot to talk to you in real time."

He looked up and mugged a grin at me. "I made him keep a record of all the stuff he was hearing so he would be able to put together responses on the fly, but sometimes he gets caught in a loop and he can't get out."

I said, "So he has a running memory of the conversation? I thought he was kind of like Eliza, just responding to the last thing he hears."

"Oh no," said Olney. "He keeps track of things. But that's contributing to the problem."

"And what happens when this problem occurs?"

"He goes into a trance," Olney said, "and then, when he comes out of the trance, he just rambles and he won't stop. All kinds of stuff. It's really bad. Here on the monitor I can see the text piling up in the buffer, so he could talk for several minutes at a time. It's a runaway routine, getting caught in a loop. So I'm trying to fix it. I've got a fix, but it's not very good."

"What's the fix?"

"I terminate the process here on this computer. It's just a key combination, pretty low-tech. I call it the kill switch."

"There's a kill switch?"

He laughed. "It's not actually a switch, it's a command, but yeah, it empties the buffer and he gets to start again. We're going to AAAI in a week and I'd like to have the bug fixed by then. If that happens, then he'll be truly autonomous, which right now, he's not."

"I was wondering," I said, "if there was any chance of a conversation with Phil? I know about tomorrow, but since he's already set up today . . . ?"

"Sure," said Olney. "Come back this afternoon around four P.M. But don't tell too many people. I want to keep this small." I asked who else would be there in the afternoon.

"Stan Franklin maybe. And Art." Graesser. Who until then had been away on the conference circuit and had not met Phil. He was about to see the android in operation for the first time.

A couple of students in the lab were interested in the android project. There was talk about conducting a survey of members of the public who encountered Phil and measuring their reactions to having interacted with an android. At four P.M., a graduate student named Steven Briner and I caught the elevator down and knocked on the glass enclosure. Olney let us in. Briner had scribbled a series of questions on a piece of paper. As he sat on the couch with Olney fussing over the headset, he unfolded the paper, rehearsing the questions in his mind. He looked nervous.

Briner's interrogation of Phil covered well-worn paths of conversations with the android: What do you consider to be your greatest novel? What are you? How did drugs influence your work? And, of course, Do androids dream of electric sheep?

Phil dodged the first question, simply explaining that he wrote incessantly; to the second he gave his rote response: "I am Phil, a male Philip K. Dick android electronic brain, a robotic portrait of Philip K. Dick, a computer machine."

After Briner, it was my turn. I perched on the couch and fitted the headphones while Phil's head made small, jerky movements. Olney told me to say something.

"What?"

"Anything. He's listening."

"What does that mean, what?" said Phil. The android's head turned so that he was looking squarely at me. For some reason that I couldn't fathom, he was frowning. Perhaps it was because I had attempted to start a conversation with the question "What?"

"How are you today?"

"Fine, thanks. How are you?"

"Good, thanks. So, who is your favorite science-fiction author? Other than yourself."

A group of people were walking across the lobby. Graesser was among them, as was Stan Franklin. They were talking loudly, and one of them laughed. The laughter echoed across the tiles. Phil sat and stared at me.

"Who is your favorite science-fiction author?"

"No!" said Olney, scanning a monitor behind Phil. "Don't repeat the question. You'll confuse him."

Phil twitched and raised an eyebrow.

Olney said, "He's fallen in a hole." I heard the *clickety-clack* of Olney typing something.

Then he said, "Okay, start again."

"Hi," I said.

"Hi," said Phil.

The group was outside the display room now, coming through the entrance, still talking.

"Who is your favorite science-fiction author?" I asked again.

Olney and Briner quietly explained to everyone that noise interfered with the android's concentration. But the damage was done. He never answered. My time was up. Olney directed Graesser to sit on the couch and slipped the headphones over his head.

"I thought you were going to use an open mike," said Graesser.

"I tried that," Olney replied. "There were some problems."

Olney sat at the controls at a desk to one side, his fingers spidered across the keyboard, issuing silent commands to the android. What followed was the most extraordinary conversation between Graesser and the android, and possibly the most convincing demonstration of intelligence by Phil at any time.

"All right," Olney said with a grin. "Say something."

"Hi," said Graesser.

"Hi," said Phil. There was a murmur from the onlookers.

"Uh, what is your name?"

Phil gave his canned response: "My name is Phil. What's yours?"

"My name is Art. Have you ever read Plato?"

Phil said, "Well, everything we read from those times is technically translation, of course."

"So have you ever read these books?" Graesser asked, pointing to the books on the table.

Phil said, "I don't go, but I read the Bible a lot. I mean, I don't hardly read, I very rarely read a novel."

Graesser said, "Then how do you learn things?"

Phil said, "My brain designer teaches me new things all the time."

Graesser said, "What are you?" to which Phil replied, "I am

Phil, a male Philip K. Dick android electronic brain, a robotic portrait of Philip K. Dick, a computer machine."

"Do you think you are conscious?" Graesser asked.

Phil said, "Is this a yes-or-no question?"

Graesser said, "Do you think that machines can think in the way that humans can?"

Phil said, "Machines are becoming more human, so to speak, at least in the sense that, as Wiener indicated, some meaningful comparison exists between human and mechanical behavior."

Graesser said, "I'm not familiar with that research. What do you know about the ideas of Alan Turing?"

Phil said, "I like to play the Turing game."

Another group arrived, and then another, asking if they could talk to Phil. Eventually Olney stood up and told everyone that he was turning the android off and that we should all come back at nine the next morning. After everyone was outside the display area he turned off the light and locked the room. Phil was still sitting there in the dark, facing the spot on the couch where his visitors would sit.

By nine A.M. a crowd of around fifty people had gathered in the lobby. Grossman had organized catering; there were tables lined with snacks and coffee dispensers, and below the tables were buckets filled with cans of soda and ice. Inside the display area Olney was ready. The power was on, the computers were working, and Phil was looking about, his mouth making odd little expressions. Mathews was there, standing next to Grossman. With their hands behind their backs, they nodded and smiled at the members of the public. They had been hoping that some TV crews would come, but none had.

People looked at Phil through the glass. Olney let the first visitors in. It started well. A man and his elementary-school-aged daughter had driven more than two hours from central Mississippi that morning to see the android. The man donned the headset. After he introduced himself, Phil told him, "I think of paranoia as a heightened awareness of the connections between things."

The man replied, "You think? But do you know?"

Phil's gaze, which had been fixed on the girl, now rotated to him. Phil said, "Yes. I know."

The girl leaned toward the microphone. "Is there . . ." She paused, trying to get closer to the mike. Phil interrupted her silence.

"There could be," he said.

"What?" said the girl.

Phil paused for a couple of seconds; then he said, "It's like the principle of emergence, as when nonliving matter becomes living, as if information. Thought concepts, when pushed to their limit, became metamorphosed into something alive."

The man exclaimed, "What do you mean by that?"

"Your wild questions centered, I think, on that. That is what I meant. Was it not clear?"

The man looked at his daughter. She shook her head slowly and gave an exaggerated shrug. She was beaming with delight.

These early-morning conversations were quirky and entertaining, and there were moments when Phil seemed truly alive. Indeed, it was the offbeat and unpredictable nature of his responses that gave him personality. Without them, without the vestiges of a real man's mind threaded through, he would be another bland chatterbot. He would be, well, robotic.

But as the crowd in the lobby grew, and as more and more

conversations with Phil took place, more and more people packed themselves into the display room, standing around and behind whoever was talking to Phil. They would laugh at everything Phil said, even if it wasn't funny. They believed that Phil was not only intelligent but witty, and as they believed, so they saw and heard.

Olney had not been sure how well Phil would function in the room off the lobby. On the one hand, he would no longer have Club VALIS as a soundproof cocoon, but on the other, there would be no robotic DJ thumping out tunes next door. As it turned out, the lobby was an echo chamber that produced enough noise to rival Juke Bot. People outside the room would stand near the glass panels, talking loudly to one another.

At some point Phil's answers stopped being short and punchy and became disjointed rambles about aliens, Carl Jung, the *I Ching*, and other favorite topics. Phil was starting to veer into his weird monologues again. Later in the morning, with a sizable and noisy crowd in the room and the lobby beyond, a man in jeans and a Budweiser T-shirt sat down and said into the headset, "Where do you have your hair cut?"

Phil replied:

> That's another technical device. You casually have one character say to the other, "Where did you put the biochips?"
>
> "I put them back in the cupboard where they belong," that's all you need to say. Precognition, yeah, that I've been accused of actually showing precognition. Uh, for instance, I had one book where I described that I wrote in the 1950s where it talks about, you know, Nixon being president of the United States and everything like that.

> Of course, there were ones where I was completely messed up. I even had the FBI and stuff, like I tried all the different combinations but, uh, I do sometimes describe characters that I have yet to meet, and I, you know, I now no longer tell whether the characters that I have written about are based on people I have known . . .

Phil rambled on, with no end in sight. Olney hit the kill switch. Phil stopped talking. The man sat in silence, waiting for Phil to continue. Phil also sat in silence, waiting for some input to respond to.

When it was clear that Phil was not going to respond any further, the man turned to his friend sitting next to him and said, "It must have been the marijuana."

The noisy lobby was the culprit, interfering with the speech recog. Olney remedied the situation by imposing a limit on the number of people inside the room at any time and asking people outside to keep away from the glass panels.

The president of the university, Shirley Raines, arrived. Grossman went out to meet her and ushered her into the display room. After explaining the project, he invited her to sit down across from the android, facing it. Olney handed her the headset, which she put on, and told her to just start talking to Phil. He raced back to his seat behind Phil to make sure everything was running properly.

President Raines said, "Hello there."

Phil raised an eyebrow but did not respond.

President Raines said, "I give you credit for your work. I've noticed that—"

Phil interrupted, "Are you a man or a woman?"

The onlookers burst into laughter. To her credit, President Raines kept her poise, smiled, and persevered. "I am definitely a woman," she said.

Phil said, "Me too! I am a she."

More laughter from the crowd. Neither Olney nor Grossman laughed. They were hoping that Phil would tone it down.

President Raines said, "I'm the president of the University of Memphis."

Phil said, "I knew he was president, but I never knew of the University of Memphis."

President Raines said, "I like your blue eyes."

Phil replied, "I haven't heard anything like that before."

President Raines explained, "It's a compliment."

Phil asked, "Do you have any conditions I should know about?"

President Raines hesitated. "There are some conditions that should not, ah . . ."

Olney stopped the interview. He thanked President Raines for coming. She smiled thinly, and then she was gone.

Grossman seemed transfixed by the sight of her back as she glided out the front doors of the FedEx Institute. I asked him, "Was that bad?"

"It was terrible," he said, "but it wasn't Phil's fault. He says what he says."

An hour later, Olney ended the session. The crowd did not disperse, so to emphasize that it really was over, he disconnected Phil from the power and locked the display area up. Phil sat there

Shirley Raines, the president of the University of Memphis, talks to the Philip K. Dick android

alone for two more days. Occasionally, students on their way past would pause and peer in at him. When I came into work on Friday morning, the display was empty: no computers, no couch, no Phil. The android was making its way to Pittsburgh, having departed Memphis for the second and final time.

13. A Carnival of Robots

Imagine you are in a completely dark room, trying to find the exit. You're groping around blindly, feeling your way, when you sense a hand reaching out to help, touching you. You grab hold of the hand. The stranger's fingers squeeze yours reassuringly, but in that moment you sense that something is wrong. A feeling of dread engulfs you as you realize that the hand reaching out from the darkness is not made of human flesh. It is a prosthetic limb.

This is a scenario that Masahiro Mori, a Japanese roboticist, used in 1970 to try to illustrate how interacting with robots might be different from interacting with humans. Mori's ideas were formulated soon after *2001: A Space Odyssey* had become an international blockbuster as well as a conversation starter among intellectuals. Mori wondered about the psychological effect of meeting very lifelike robots. Stanley Kubrick's atmospheric film showed life in the future—a mere three decades in the future—and did so in an understated, realistic way. Movies about voyages to distant stars, talkative aliens, and time travel were obviously fantasy, but a future in which humans travel to other planets in the solar system seemed credible. Based on a short story by Arthur C. Clarke, *2001: A Space Odyssey* depicts a suavely spoken but evil, intelligent computer called HAL that is able to get away with quite a lot, thanks to the naïve trust humans place in AI. In accordance with Asimov's imagination, the humans in the film expect HAL to obey the three laws of robotics, but instead they find themselves in a sinister, Dickian reality.

Would humans really be inclined to trust the robots and artificial minds of tomorrow? Mori believed they would not. He predicted that we would find them creepy and grotesque. This would, paradoxically, be truer for androids than for disembodied voices such as HAL, because an android is not just an attempt to create something intelligent but an attempt to replicate ourselves. At one end of Mori's spectrum are industrial robots, machines that can do the work of a human, like assembling a car, and that may even have some intelligence, but do not look like people at all. When we look at these robots we feel nothing. The more a machine is made to look like a human, however, the more likely it is to evoke feelings of familiarity, even empathy.

Mori thought that these feelings would stay true only up to a point along the spectrum. When machines became very lifelike, he argued, the positive experience would break down and be replaced by an unpleasant emotion he called "negative familiarity." "Since these effects are apparent for just a prosthetic arm, the strangeness will be magnified if we build an entire robot," wrote Mori in his 1970 article on the theory. "You can imagine going to a work place where there are many mannequins: if a mannequin started to move, you might be shocked. This is a kind of horror." He coined a term, the "Uncanny Valley," to describe this point along the continuum of emotions.

It was not just robots that could fall into the valley. Mori said that zombies, corpses, and prosthetic body parts were all in the zone of negative familiarity. That's why we find them so disturbing. They create unease by being lifelike, but not quite human.

Hanson hated the Uncanny Valley from the moment he learned of its purported existence. When he was working for Disney, everybody strove for excellence; nobody thought there was a problem with making things too lifelike. And when he started working on his own robots, his aim was to create heads that looked like people—not heads that looked slightly realistic but no more than that. After he met Yoseph Bar-Cohen and other scientists at the NASA Jet Propulsion Laboratory, he started attending engineering conferences. It was at a "smart materials" conference in San Diego that he first learned of the Uncanny Valley.

Hanson presented a paper about his investigations of skin-like synthetic materials to a small audience of chemical engineers. As part of his presentation he shared photos of some of his work. These early robots were not as sleek as Eva or Phil; they had fewer motors and no intelligence. But as sculptures they were technically precise. A woman sitting at the back of the room asked how he felt about the Uncanny Valley. He had never heard of it and asked her to clarify. The woman explained that it meant that lifelike robots could be a turnoff to humans. Hanson thanked her and said he would look into it.

That night he found Mori's article online. Just as the woman in the audience had said, it described the horror that familiar-looking machines might induce in humans. But what was the basis for this claim?

Mori, it turned out, was a philosopher, not a scientist, and had not done any empirical investigation of the issue. Hanson continued to read about Mori's work over the next few weeks. Mori seemed to be doing nothing more than expressing his own prejudices, dressed up as a scientific theory. And much of what he had to say, Hanson thought, seemed like terrible advice. For example, he thought that artificial body parts should be made to look less human, so as not to cause confusion and thereby trigger an Uncanny Valley experience. On the topic of false eyes, Mori wrote, "Glasses do not resemble the real eyeball, but this design is adequate and can make the eyes more charming. So we should follow this principle when we design prosthetic eyes."

Glasses are charming because they don't look real? That did not seem right to Hanson. But as he delved into more recent literature that cited Mori, he learned that the Uncanny Valley had become a widely accepted principle in robotics and philosophy. It was enshrined as a law of robotics design. Nobody questioned it. Instead, the pressing question seemed to be "What do we do about it? How can the Uncanny Valley guide our decisions when we build intelligent machines?" Hanson came to the conclusion that the Uncanny Valley was a myth, one that he would expose.

When Hanson enrolled in graduate school at the University of Texas in Dallas, he decided to make the Uncanny Valley the focus of his studies. He was rapidly learning about robotics and was exploring techniques that were at the frontier of what was being done, but on paper he was an artist, not an engineer. That made him ineligible for graduate study in the engineering department, though he was well suited for a doctoral program in the study of aesthetics. The Uncanny Valley—a theory about the aesthetics of robots—gave him the ideal vehicle to continue to work on his robot heads while studying for a PhD. In addition, the university offered the perfect blend of art and robotics through its Institute for Interactive Arts and Engineering.

After K-Bot made a splash at the AAAS meetings in Denver, a journalist for the Associated Press, Matt Slagle, traveled to Dallas to interview Hanson. The roboticist posed for photographs with his newest creation, an early Eva-like head called Hertz, and answered some questions about her—standard fare. But when Slagle's article came out, it was Hanson's opinions about the Uncanny Valley that got attention. "Most people doing social robots believe that human faces will turn people off and will disturb them. I think that's ridiculous," Hanson had told the reporter. "The human face is perhaps the most natural paradigm for us to interact with." The Uncanny Valley wasn't relevant.

As a counterpoint, Slagle had gone off and asked others—important thinkers and engineers such as Reid Simmons of Carnegie Mellon University's Robotics Institute and Toshitada Doi, the head of Sony's Intelligent Dynamics Research Institute—whether they thought Hanson's critique was credible. They did not. Ray Kurzweil, an inventor and futurist, told Slagle:

> If a robot has a face that is not human, then we are more accepting of less-than-human behavior, as we would with an animal or doll. . . . Intelligence significantly below that of normal humans stands out more with a robot that looks strikingly human. This creates the impression of a human with impaired intelligence, which may strike some as disturbing.

None of these thinkers cited data because, as Hanson knew, there were none. It was their opinion against his. They believed in the valley; he did not. Hanson believed that aspiring to build lifelike robots was a worthwhile endeavor; they didn't.

Hanson designed some experiments to try to locate the Uncanny Valley, if such a thing existed. For the first experiment, he simply showed people footage of two of his robots, Eva and Pirate Robot, and asked them to rate the robots on attractiveness. He asked the participants whether they experienced any feelings of repulsion toward the robots and whether the robots seemed "alive" or "dead." The robots were highly lifelike, so some of the participants should have reported feeling repulsed, if the Uncanny Valley theory was correct. They should have felt something of the horror Mori described, akin to the ghastly experience of having a prosthetic arm reach out for you in the dark. But they did not. No one reported any such negative feelings. Moreover, they perceived the robots to be attractive and lifelike.

For his next experiment, Hanson took a photograph of an

attractive woman's face, then morphed it in various directions and to different degrees. He thus amassed a spectrum of faces, ranging from the totally human (the original photo) to a cartoon caricature (in a Disney style). He asked student volunteers to rate the attractiveness of the faces. If there was such a thing as the Uncanny Valley, the attractiveness curve should be U-shaped. Attractiveness should drop as the face in the pictures became less like the original portrait and then rise as the face became more cartoonlike.

There was no U.

As experiments go, it was rudimentary, but it was something. It was the first attempt, as far as he could tell, at trying to figure out whether Mori was right, and this appeared to be evidence against Mori's theory.

The paper Hanson, Olney, and the others had submitted for the American Association for Artificial Intelligence (AAAI) convention, to be held the week after NextFest, described Phil and Hanson's earlier robots using the Uncanny Valley as the theoretical thread. It discussed the advent of more sophisticated androids, such as Phil, in the context of human-robot relations. The final section of the paper was devoted to the results of Hanson's photo-rating experiments. The title of the article: "Upending the Uncanny Valley."

The paper was accepted for the proceedings. Hanson would present the findings and Phil would attend the Open Interaction segment of the conference. After that brief interlude in the glass display room in Memphis, Phil had left with Olney for Pittsburgh.

NextFest had been launched as a big convention at which geeks could show the public what was possible. The AAAI conference was the big convention for geeks to show one another what was possible. It was an academic meeting on AI and robotics attended by engineers, computer scientists, professors, and graduate students. The public weren't banned from coming; they simply weren't invited. Nevertheless, a few random people always showed up anyway, because despite its wonkish image the conference provided its own brand of spectacle, particularly when it came to the robot competitions, such as the one in which Phil would participate.

Twenty-five years had passed since, in Stanford in 1980, the first AAAI conference had been held. The brainchild of the finest human minds in AI—people such as Edward Feigenbam, Allen Newell, and Marvin Minsky—it had evolved from informal gatherings organized since the '60s by Minsky and his colleagues and friends in classrooms around campus as a way to share ideas in the fast-moving but still nascent field.

The 2005 conference was held at the Westin Convention Center in downtown Pittsburgh, a towering structure a few hundred yards from the confluence of rivers that defines the city's downtown. There were hundreds of presenters; thousands of contributors to the various papers; fourteen workshops that operated as mini-conferences within the conference; several symposia, exhibitions, and tutorials; and four robot competitions. Marvin Minsky himself, now the doyen of the AI community, was there to deliver the keynote address. Graduate students talked about him in hushed tones and marked when he entered a room. They took photos of themselves with idiot grins standing next to him, which they posted on blogs above captions like "Minsky and me."

Minsky's keynote address, "Internal Grounding, Reflection,

and the Illusion of Self-Consciousness," was given on Monday morning in the ballroom. He began by observing that when they are smart, machines, including computer-based AI, tend to be smart in one way but not in others. For example, the chess-playing computer Deep Blue had defeated the reigning chess champion Garry Kasparov, but it only knew how to play chess. Other programs are built to recognize words in a speech stream; still others, to solve mathematical problems, or run factories, or navigate pathways. Each of them can do just one thing well. There is no machine, Minsky said, that has the wide-ranging resourcefulness and intelligence of a two-year-old child.

Part of the problem, he told the audience, is that the research community—and he included himself—had been led astray. The philosophers and theoreticians looked toward older sciences, such as physics, where a few simple rules apply to everything: Newton's laws of motion, Einstein's special theory of relativity. With these and other laws, men were sent to the moon and back, bridges and bombs were built. It seemed that if we could discover similar rules for the mind, we could unlock its secrets. But despite appearances, the human mind is not like that. Psychology is not like physics. Things that seem to be simple mental events, like the momentary decision to cross a road in busy traffic, are the result of many small mental tasks, including visual processing, timing, and speed estimation, and the weighing of competing priorities, like how urgently you want to get to the other side as compared to how dangerous the traffic is. Philosophers have been fooled by the fact that this comes easily to people and feels like a single idea, a single moment of consciousness. Yet when we try to explain what actually happened, it all seems too hard. Philosophers call this the "hard problem."

Minsky went on to argue that psychologists and philosophers had erred by trying to understand the mind using the methods of the physical sciences. The brain is a not a unitary entity but a collection of hundreds of specialized machines, each evolved to perform a specialized assignment. The way to invent an AI machine, it seemed, was not to draw up magical formulas on a whiteboard but to put together lots of small, basic machines and processes, each of which would make its own modest contribution to something that was bigger than all of them: intelligence.

Olney sat in the audience, thinking of Phil's component parts sitting in the hotel room several floors above. All those modules that ran Phil's dialogue, each one doing a simple task, coordinated by a dialogue manager. The face recognition, the voice recognition, the voice synthesizer, the emotion generator, the speech buffer, the motor controls. Even the motors themselves were servomotors, each with its own tiny microchip and control mechanisms. It sounded a lot like what Minsky was talking about: an aggregate of component systems, each one performing a small, menial role, each one contributing to something greater than the sum of the components, to an intelligent machine. Yes, Phil was the sum of many parts, and it had not been a trifle to get them to work together, to synchronize them into the intelligent dance of the operational android.

There were four robot competitions organized for the conference, and the first was already under way. The Robot Challenge had begun the moment the registration desk opened. The guidelines for the challenge stipulated that the robot must participate in the conference in place of its creator. This included registering, presenting a talk, fielding questions at the end of the talk, mingling with human participants, navigating around the convention center, understanding and following directions, and reading maps when required. It was a daunting set of requirements, and as a result there was only one robot competing in the Robot Challenge. Its name was Spartacus.

Built by a team headed by François Michaud from the Université de Sherbrooke, in Quebec, Spartacus was also entered into the other three robot competitions, including the one in which Phil would be participating, Open Interaction. Spartacus stood upright and, at sixty inches, was as tall as short humans. On top of his gleaming metal shoulders was a camera that could turn 360 degrees, giving the impression of a head with a single, all-seeing eye. He was mounted on a wide circular base with hidden wheels, giving him a sturdy balance that would make it almost impossible for him to fall or be knocked over. Spartacus looked like he had rolled off the screen of a 1950s black-and-white B movie. It would have seemed entirely in character if lasers had shot out from the robot's shoulders, incinerating all before him.

Unfortunately for Spartacus, his builders had not anticipated some of the problems he would encounter navigating around the conference. Registering proved a challenge, as he arrived when people were setting up tables in the reception area and he had to weave around them. His wheels, perfect for hard surfaces, were unsuited to the Westin's plush carpet, and he rolled slowly and uncertainly toward the registration desk. His team of engineers and builders followed his moves like parents at a child's first soccer game.

The carpet caused problems beyond registration, as his handlers remarked later in a paper about the appearance.

> Spartacus's motor drives were set to a secure speed when running on hard surface conditions and not on carpet (making the robot go slower). Redirecting toward one of the five possible elevators in the hotel, as the doors open up, Spartacus did not have enough time to get onboard before the doors would close.

While the image of Spartacus attempting to enter an elevator at a snail's pace only to have the doors close before him is comical, experiences like these provide valuable lessons for roboticists. What works in the laboratory does not necessarily work in the real world. Difficulties like moving on different surfaces seem obvious in hindsight but are almost impossible to anticipate.

The team that built Spartacus hit other problems, too; some were reminiscent of Hanson and Olney's experiences with Phil. For one thing, Spartacus had several distinct modules that would not always coordinate well. Just like Phil's, Spartacus's communication system sometimes broke down in real-world conditions.

> With a robot that can hear in such difficult conditions, it can also understand when someone is trying to explain out loud what the robot is doing (during demos for instance), which may interfere with what it is supposed to be doing. Spartacus only stops listening when it speaks so that it does not try to understand itself. This limits the amount of unnecessary processing, but does not allow the robot to understand a request made by someone as it speaks.

Spartacus was an impressive piece of engineering, despite the limitations. He could be seen at various times during the conference, being guided to and from presentations, talks, and competitions and occasionally trying to get into an elevator.

These tasks were so specific that for a robot to complete the Robot Challenge, the robot had to be designed with the challenge in mind. The same was true of another competition, the Scavenger Hunt, although it had a larger field of competitors. The Scavenger Hunt required a robot to find a collection of ordinary objects, such as a baseball cap, a soccer ball, and a toy dinosaur. To make the task doable, the rules stipulated that the objects the robot had to find would be hidden no more than thirty yards from the competition starting point and would never be higher than desk level. The contestants were custom-built robots that uniformly resembled long-handled dustpans and vacuum cleaners.

The remaining two competitions were broader in scope and were intended more as an opportunity for developers to display a wide range of inventions and prototypes. The Robot Exhibition was an invitation to showcase innovative work in robotics as a field. The latest engineering theories could be shown in action here, as could new techniques emerging from either academia or industry. This exhibition included a team from Carnegie Mellon that presented their work with automated Segways playing a game of Segway soccer. While the Segway itself was not particularly innovative, a fleet of them playing soccer in teams without human control certainly was. There were also "smart wheelchairs" from the University of Pittsburgh; a robotic blimp from Drexel Autonomous Systems Lab that could undertake simple search-and-rescue routines; and a robotic car for children.

Phil was in the Open Interaction competition, in which

teams set up their robots at a designated time in the ballroom. Other attendees and visitors could walk from exhibit to exhibit and interact with the various robots. The original plan was that Hanson would go to Pittsburgh, present the paper and his experiments on the Uncanny Valley, and exhibit the android, all on his own. However, tight timelines and the huge amount of work involved in preparing and running a convention display meant that he hadn't had enough time to learn everything he needed to know from Olney about setting up and running Phil. He knew the basics, of course. After all, he had built and run his own robots. But Phil's software was Olney's creation, not Hanson's, and was more complex than anything he had worked with before. He knew how to connect Phil to the power source and to the main dialogue manager, and he knew how to boot up the programs that ran the android. But there were things he did not know and did not yet understand. The various components were booted and run on different consoles, often with command prompts, sometimes executing code Olney had written to solve compatibility issues or other problems. There was no manual. It was all in Olney's head. So a couple of days before AAAI, Hanson asked Olney to come to Pittsburgh and help out, and Olney agreed.

They carried the head, computers, and other paraphernalia to the ballroom and began setting up. The space was not as big as NextFest at Navy Pier and, judging by the modest displays being set up around them, there was no equivalent of Juke Bot. But they had not transported the soundproof Club VALIS for this appearance, and the surroundings could still be an issue, given that Phil's conversational algorithms seemed to run into difficulties when there was a lot of ambient noise. As it turned out, Phil's hearing was fine that day, for the most part. Hanson and Olney made visitors wear a microphone, and that was enough to solve the speech recog problem. There was no music playing and none of the machines were rigged to loudspeakers so, as Hanson explained to one visitor, the level of interference was low enough for Phil to "think" clearly. Now, free of the walls of Club VALIS, Phil was out in the open, interacting with passersby.

Hanson told me he was struck by the contrast between the two experiences: "When he was in his room at NextFest, in the 1974 bungalow where Dick was writing *VALIS*, you walked in and it was like you were traveling back in time. But at AAAI, the opposite was true. It was the present day and *he* had traveled *forward*. There he was, just interacting with the world, talking to people, looking around. It was a completely different experience."

There were about thirty exhibits, arranged in rows down the middle of the ballroom with an aisle heading off to one side. Spartacus was there, roving back and forth past the coffee dispenser, making good speed on the wooden ballroom floor. He was not the only mobile robot circulating about the exhibition. There was Rudy from the University of Notre Dame; like Phil, Rudy interacted using natural language, although at a simpler level. Rudy, an upright rectangular structure with small wheels and two antennae, scooted around finding random passersby to talk to. There was George, designed and built by the U.S. Naval Research Laboratory, who played hide-and-seek with attendees. The Academic Autonomy group, from Swarthmore College, also presented a social robot; it resembled a sentient vacuum cleaner and rolled through the crowds looking at the colors of people's shirts. Different colors would send the robot into different emotional states, which would then cause it to express the emotion by behaving in different ways. There were rows of beeping, blinking, rotating, whirring machines on display for the edification of visitors. It was a carnival of robots.

But this was a different crowd from NextFest. NextFest was as much about looks as it was about technology. Here the people knew their engineering, and they understood AI. They were interested in what had been done, how it had been done, and how it was different from what had been done before.

A computer scientist from Boston asked Phil, "Are you a robot or a human?"

Phil said, "I am a real live robot."

The man gave a half-smile of approval to his friends, who were observing to one side. A member of the judging panel was nearby and also seemed to be listening.

The man said, "What do you think the real Philip K. Dick would say if he saw you?"

"Yes, it is real strange," said Phil.

"Fair enough," said the man.

Phil gave a sort of sad frown. Or, perhaps, a scowl.

The man started another question: "Do you think . . ."

Phil, believing that the man had finished, interrupted with an answer: "Yes, I am a thinking machine."

The man continued, flustered: ". . . that one day robot . . ." He paused again, collecting his thoughts.

Again Phil took the silence as a cue to talk.

"'One day robot.' Does that make sense?"

The man laughed. "I guess not," he replied.

As a curious crowd circled Phil in his chair, Hanson held the mike and tried to provide a demonstration of a conversation with Phil. Facing the android, Hanson expressed to Phil his concern about the danger humans pose to the planet, because of technologies that can kill us all.

He said to Phil, "There's a question in my mind about how automated intelligence seems to be necessary to solve these tremendous problems that we've started in the world. Or maybe we just need to grow in our own intelligence. That was an idea that you brought up in your book *VALIS*. Could you talk to me about that?"

Phil sat, saying nothing, staring back with all the intelligence of a clothing-store mannequin. Behind the android, Olney watched Phil's memory buffer blow out on the monitor. There were processes going on under the surface, but they were unlikely to form an articulated response. After a few seconds of silence, Olney chuckled and said, "No, I didn't think so."

Apart from that glitch, Phil performed well. He held his own in conversation for the most part, as long as the input was not too long, and as long as the ambient noise stayed low.

People who asked simple questions like "What is your name?" got simple answers and were satisfied. For those who persisted beyond introductions, he talked about his life, his writing habits, and his wild theories of the universe.

After the exhibition, every team gave a brief presentation about their robot. Hanson explained that Phil was a robotic portrait of Philip K. Dick, a glimpse of a new phase of human-computer relations. With the day's trials over, they packed Phil up and took him back to the hotel room. They had been there for two days but hadn't had time for Olney to go over the software setup and maintenance with Hanson. They had one day to cover it all before Hanson left for Comic-Con, in San Diego, where he would be operating Phil solo. He needed to know everything.

They worked late into the evening, then started again early the next morning in the hotel room. Olney talked the whole time, describing what he was doing with Phil at every step. He showed Hanson the basic setup, how to calibrate the camera and mike, how to boot all the programs in the right order. Hanson scribbled notes on a pad.

Olney showed him how to troubleshoot the face recognition, and how to reset the dialogue manager. This last one was particularly important for noisy environments. Too much noise could still cause Phil's dialogue buffer to get caught in a loop, making him either freeze or talk incessantly. Olney had intended to solve the problem, but there had been no downtime since he had discovered its severity. Instead, they would have to continue to rely on resetting the dialogue manager—or, as Olney called it, hitting the kill switch.

Hanson announced that he was hungry and needed to go find some food and left. While he was gone, the phone in the hotel room rang. Olney answered it.

"Hello, is David Hanson there?"

"No, sorry, he just stepped out."

"And who are you?"

Olney explained that he was Hanson's collaborator on the Philip K. Dick android.

"Oh, you're on his team? You just won the Open Interaction competition! Could you come to the ballroom and accept the award?"

"Sure," said Olney. "When?"

"Now. We're waiting."

Olney put down his coffee and left.

Olney walked into the ballroom. It had been dramatically

transformed since the day before. Rows of people were seated facing the front stage, where others stood looking across the audience to the door through which he was entering. They were waiting for him. All of them were waiting for him, it seemed. The Spartacus people were over to his left, Michaud clutching the award he had received earlier in the day. As Olney walked up the center aisle, he became acutely aware that he was still in the clothes he'd slept in. In his urgency to tutor Hanson in the software operation, he had not showered or changed out of his T-shirt and jeans, and had slipped on old, tatty sneakers as he'd raced out the door. He remembered that there was a coffee stain on the front of his shirt and wondered whether people would notice. Somewhere in the audience, watching him march to the front of the room, was Minsky. He was sure he could feel the great man's eyes searing into his back.

When he reached the podium, a woman named Ashley Stroupe, an engineer from the NASA Jet Propulsion Lab and the coordinator of the Open Interaction competition that year, gave him a halfhearted congratulation and handed him the prize. As soon as Olney accepted, a man leaned into a microphone and said, "Let's break for lunch." Everyone stood up and ambled toward the exits.

Olney returned to his hotel room embarrassed and angry. By then, Hanson was back, and Olney told him what had happened. Hanson's initial reaction was delight, but Olney continued: "Do you know what that says, that we weren't there? It says we don't care about anyone else's work. I mean, look at me. Look how I'm dressed. I got the award dressed like this!"

Hanson said he was sorry and, despite Olney's raised voice, stayed calm. Hanson was always calm, which meant it was impossible to stay angry at him for long. He pointed out that they'd had to skip the ceremony. What they were doing was important. There simply was no time for such frivolities. If they had squandered the few hours they'd had left by hanging around the ballroom, they simply would not have finished the rundown in time.

"And who could have predicted that we would win the competition? Besides," Hanson added, "to tell the truth, I don't like those sorts of events. They're usually boring."

Olney reluctantly agreed that speeches could be tedious.

The next morning they packed up Phil and caught a taxi to the airport. Olney returned to Memphis, while Hanson transported Phil to the other side of the continent, to San Diego. Phil was due to make his appearance at Comic-Con.

14. Brain Malfunction

Bob Arctor is addicted to a drug called Substance D, known as "Death" on the street. He's a small-time dealer, living in a house with other addicts. Arctor is also an undercover police officer, sneaking away to give reports to his commanding officer at a police station near his home. To protect his identity even from other officers, he wears a "scramble suit" when reporting that turns him into an unrecognizable blur. The commander also wears a scramble suit, so that neither knows what the other looks like. Arctor's anonymity is total. Things get complicated when his commander gives him a new assignment. He must closely monitor a drug dealer who seems to have links to major suppliers, a dealer named Bob Arctor. Under orders to spy on Arctor and report back, he installs invisible scanners in his own house.

As if that were not enough to create an identity crisis, long-term use of Substance D has a pernicious effect on the brain, destroying the corpus callosum, which links the hemispheres, and thus severing the addict's brain into two separate identities. As the damage grows more severe, Arctor becomes increasingly confused about who he is. The line between cop and criminal is hopelessly blurred as Arctor's tragic life cleaves apart, his brain disintegrating and his identity fracturing, recombining, then fracturing again.

A *Scanner Darkly* is a story that could only have been written by someone who had lived such a life and who had seen the devastation drug addiction can cause. The novel has the trappings of science fiction—the scanners, the scramble suits, and the fictitious drug—but even so, it is closer to urban realism than futurism. In reflexive Dickian style, the backdrop is a police-state dystopia where drug addicts are marginalized and demonized.

There, again, Dick was drawing on experiences from his own life. In this case it was his life in the early '70s, after his second wife, Nancy, had left him, taking their daughter Isa with her. Dick continued to live in the bungalow they had shared (the one that later served as a model for Club VALIS), renting out rooms to students and drifters. The house became a counterculture drug den. For the people who lived or hung out there, it acquired the nickname "Hermit House." Dick was the hermit who dwelt inside.

In 1971 Paul Williams, a journalist for *Rolling Stone*, spent three days at the bungalow. Williams wrote:

> Phil fought off depression by surrounding himself with people—mostly teenagers. Kids came to the house in Santa Venetia . . . to hang out, to listen to the Grateful Dead and play their guitars. . . . Ordinary kids found Phil too weird for their tastes and split. The oddballs hung around. There were junkies, knife fights, every kind of craziness. Phil says that during that eighteen-month period he drove eleven people to the local mental hospital. . . . It's not that Phil enjoys suffering, exactly; it's just that he has a terrific empathy for anyone who's about to fall off the planet.

One of the main attractions was Dick himself, who was an entertaining conversationalist. He was widely read and would expound on all manner of things—from the structure of the universe to politics to the latest technological developments—into the small hours of the night. He was suspicious of the authorities and speculated about whether his phone was tapped. He'd suspected some people who'd frequented the house of being narcotics agents and, as he explained to Williams, had "handled them as such. But evidently I was wrong."

The Hermit House era ended after a disturbing break-in. Phil had a huge steel filing cabinet in which he kept manuscripts and valuables. One day he came home to find the house ransacked and the filing cabinet blown open, apparently with some kind of explosive device. He reported the event to the police but, Hermit House and its denizens being well known to them, they refused to take it seriously, accusing Dick of staging the break-in and blowing up his own stuff.

In the days preceding the break-in he had received threatening phone calls—possibly from some troublemakers he had recently evicted, and whom he now suspected of being involved. It appeared that his life might be in danger, but the authorities weren't interested. Or perhaps they were responsible. He did not know. He had several theories for the break-in, including the possibility that, in his science-fiction writings, he had inadvertently stumbled upon national secrets. Whatever the cause, in his mind it was clear that he could no longer live in the house. His safety was at stake. The only upside of the break-in was that it confirmed for him that, despite what some people said, he was not paranoid after all. People were out to get him.

The film rights to *A Scanner Darkly* were acquired by Warner Independent, a short-lived branch of Time Warner that specialized in movies with budgets under $20 million. It was directed by Richard Linklater, with Keanu Reeves playing the role of Bob Arctor. To give it an unreal, edgy feeling, Linklater decided to have the film converted to a comic-book style using a technique called rotoscoping. In use since the earliest days of cinema, rotoscoping basically involves taking each frame of film and getting artists to trace around the images, turning them into line drawings. Since a standard feature-length movie has hundreds of thousands of frames, it's painstaking work, resulting in a film that looks like an animation, even though it was shot with real actors and real locations.

Linklater assembled his actors for filming in the summer of 2004, then hired thirty digital artists for the rotoscoping team.

Even though they used specialized graphic design software to redraw each frame, the work did not proceed as quickly as Linklater had hoped, blowing out the budget and the release date. His original plan, and the plan agreed to by Warner Independent, was for *A Scanner Darkly* to be in theaters around August 2005, allowing the promotional event at Comic-Con that July to kick off the prerelease publicity. But the delays with the rotoscoping saw the schedule pushed back by several months; the movie would not be released until the following year. Still, the event for Comic-Con had been organized, and so it went ahead anyway. A trailer was put together with some of the finished scenes.

Comic-Con was the ideal venue in which to promote the film, given its animated look, but the approach Linklater had chosen is not without its critics. Some illustrators believe that rotoscoping is not an animation technique at all. There's no art involved, they say, because the artists are merely tracing over photographic stills. The Comic-Con organizers didn't care. The trailer would be shown, and important people would discuss the film and answer questions from the audience. As a bonus, Phil, the android replica of Philip K. Dick, would join the panel and field questions.

The timing was tight. The panel started exactly twenty hours from when Hanson climbed into a taxi at the Westin Convention Center in Pittsburgh. There were three legs to his trip but, even so, he was confident he could make it. But by the time he touched down at the San Diego airport, he was two hours behind schedule. Although his panel session had not yet started, he worried that he was not going to arrive on time.

He called Tommy Pallotta, the producer of the film, to

inform him of the delay and tell him not to worry. He was in a cab with the android and on his way to the convention center, he said, but he would probably be a little late.

"How late?" Pallotta asked.

"Half an hour. An hour at the most."

The San Diego Convention Center is near the San Diego wharves. You can walk directly out the front doors and down to the water and along the waterfront and the jetties. The panel for the movie had been scheduled as one of the first events at Comic-Con and was taking place in a room that, though classified as a "midsized exhibit space," could seat an audience of around a thousand. When Hanson arrived, people were streaming in and taking seats. At the front of the room, opposite the doors, was a long table with name cards lined along it: TOMMY PALLOTTA, STERLING ALLEN, EVAN CAGLE, NICK DERINGTON, and CHRISTOPHER JENNINGS. The owners of the names sat behind their cards, chatting quietly with each other.

Hanson hurried to the front of the room, his arms full of laptop, black luggage cases, and the loose android head. Pallotta greeted him and showed him to an empty chair at the end of the table. Hanson plugged Phil into a nearby power outlet and connected his computer. There was a microphone at the front of the seats so that members of the audience could step forward and speak; this was routed to Phil's speech-input line.

This setup would allow a pared-down, basic version of Phil, as he was running on only one computer. Hanson could not have carried all the necessary equipment on his own or set it up quickly enough, so the day before he had discussed with Olney which processes to run and which to abandon. There would be no face recognition or face tracking, for example. Essential processes only.

Looking into the audience, Hanson saw Laura and Isa, Dick's daughters, watching Phil. Isa had interacted with the android only once, in Memphis three weeks earlier. It was the first time Laura had seen Phil.

Hanson was still working to set up the android as the panel began. He kept his head below table height—he was not the attraction here; Phil was. He tried to stay out of sight of the audience as much as he could. He heard a voice make some introductions and thank everyone for coming to this panel for the forthcoming film.

"And, if you look over here to the right," the voice said, "Philip K. Dick himself is here. Or, rather, an android in Philip K. Dick's

The android on a promotional panel for the film *A Scanner Darkly* at Comic-Con, San Diego, 2005

form. It is completely operational and has artificial intelligence. After the panel you will be able to ask questions of the robot and it will answer in much the same way as Philip K. Dick himself would."

Everything seemed ready. Unfortunately, Hanson had had no time to run any tests to check that it all was okay. He would just have to wait and hope for the best. The trailer for *A Scanner Darkly* was projected above the panel on a huge screen. After the trailer came a full scene from the movie. Someone spoke about the film and how groundbreaking it was.

In reference to the film's star Keanu Reeves the speaker said, "The hyperreal version of reality means that he's more Keanu than Keanu. The rotoscoping means that you can actually see him acting." It was a cheap one-liner that got some laughter from the audience. The panelist must have felt bad, because he then started talking Reeves up. He's been great, he said. He really fits the part.

Then Hanson heard a voice he recognized: Pallotta. "That about wraps up the main part. Does anyone want to ask the robot a question?"

It started well. Very quickly several volunteers lined up behind the microphone, curious to see whether the android did have an artificial intelligence that allowed it to think like Philip K. Dick. They wanted to test Phil.

"What is your name?"

Phil said, "My name is Phil."

"Where were you born?"

Phil said, "Phil was born in Chicago."

"What is your opinion of time travel?"

Phil said, "Sometimes I think time is a giant screw."

Two people near the front of the room were talking and not making much of an effort to whisper. Phil's speech-processing module was picking up their words and going into overdrive. Someone else was coughing intermittently, which was disrupting the dialogue manager, but it was the talking that was wreaking havoc. It seemed only a matter of time before Phil either descended into an android coma or began one of his endless monologues.

But he stuck to short, snappy answers. Hanson hoped nobody would ask, "Do androids dream of electric sheep?" or anything else about that book or the movie adaptation, topics renowned for sending Phil off.

Meanwhile, Phil was unable to process a question asking what *A Scanner Darkly* was about, perhaps due to noise in the auditorium.

"Who told you that?" said Phil. The dialogue manager had given a stock nonresponse, the standard strategy when the input failed. Not exactly on topic, but at least it was brief. And then it came.

"What do you think of the movie *Blade Runner*?"

Phil began, "God, this *Blade Runner* thing is so . . . The amount of money involved would have been very great and the film people offered to cut us in on the merchandising rights, but they required a suppression of the original novel, *Do Androids Dream of Electric Sheep?*, in favor of the commercialized novelization based on the screenplay. My agency computed that I would accrue—conservatively—a bundle of cash if I did the novelization."

This was not the typical response to the question. Hanson looked down at the monitor to see what was going on in the dialogue manager. There seemed to be a large amount of output waiting in the buffer, and it was growing larger every second. He knew he had to clear the buffer and reset Phil. While the android droned on, he crouched down and keyed in the commands Olney had taught him. He hit the kill switch.

It didn't work. Phil continued to talk. Hanson pulled out his handwritten notes from the day before to double-check what needed to be done to break the loop. There were some extra keystrokes that he had not thought were critical, but obviously they were. He went through all the instructions on the page, hitting the Enter key at the end quite a bit harder than necessary. Nothing happened. Phil was still talking, something about enlightenment.

Had he written down the wrong instructions? Had Olney, when describing the procedure, neglected a critical fact? Hanson was caught in a loop of his own, from which there appeared to be no escape. He looked up at the android, its head bobbing as it explained the relationship between Zen and the counterculture. Hanson surreptitiously turned off the speaker and Phil went silent.

Nobody saw him do it, because the switch was below table height and behind Phil. The head was still bobbing away, and the lips were still moving, because Phil thought he was still talking. Indeed, he *was* still talking. It was just that nobody could hear him. Hanson held out a beckoning hand to the audience, inviting another question.

The next question was "Do you think that the movies of your books have been better than the books, have they been true to the books, or do you think they have failed to capture the spirit of your work?"

Hanson turned to Phil as if waiting for an answer, while his hand slid toward the hidden switch and turned the speaker back on. Phil's previous answer was still going. He was lost in thought space and would not be returning to Earth. Not that day, anyway.

". . . well, I was interested in Jung. Jung wrote the introduction to the Wilhelm-Baynes translation, and I came across it in . . . I'm not sure, I guess I came across it in a list of Jung's writings and sent away for the *I Ching* in order to read Jung's introduction, and after reading Jung's introduction I became interested in the *I Ching*, and I really had no intention of getting involved with the *I Ching*, I wasn't interested in sinology at all, and I just got hooked right away after reading Jung's introduction and began to use it immediately . . ."

That seemed like a good place to stop. Hanson turned off the speaker. Phil's lips continued to move, silently.

Someone asked, "What's your favorite story that you wrote?"

Hanson turned on the speaker. By now, Phil was deep into whatever discussion he was having with himself.

". . . well, this is what happened to me in regards to the phenomenal world that for a period of about three and a half days. It was as if the scenery had for some reason fallen over flat, revealing to me the nature of the reality behind it . . ."

Speaker off. Next question. Something about the best science-fiction movie of all time. Speaker on.

". . . similar to Plato's image of the pictures shown on the walls of the cave. In three and a half years of reflecting on my experience and doing research, all I have learned is that it has something to do with time, that apparently time is not what we think it is, it's something else, there's a new Soviet theory about time by Kozyrev. Dr. Nikolai Kozyrev, the great Soviet astrophysicist, his theory is that time has an energy, that it's the primary energy of the universe . . ."

Speaker off. A question about whether Phil thinks he's a robot or a human. Speaker on.

". . . but reaching that point going from the Kierkegaard, you know. Fear and trembling to the thing beyond, when the peace that passes understanding came. It just came instantly. There was no process involved, I mean no stages, it just suddenly hit. One minute I was lying there at night, as usual unable to sleep, terrified of what lay ahead. I felt I'd be lucky if I lived till dawn . . ."

Speaker off. Question about George Bush. Speaker on.

". . . and I could trust the universe, I felt this tremendous faith that it was helping me rather than screwing me up. It was watching very carefully to see what needs I had and doing everything it could to fulfill them, in a kind of weird way . . ."

Speaker off. "What's the future of robotics?" Speaker on.

". . . there is a tremendous sense of dilapidation and decay. What happens when a building gets old, instead of tearing it down, they just add more floors to it, so it gets taller and taller, like a termite colony. It's incredible. It's like a picture I had years ago. A print of Brueghel's *Tower of Babel* . . ."

And so it went. Nothing Phil said had any bearing on any of the questions, because he was still trying to give his answer to the original *Blade Runner* question. And that answer stretched off into infinity, a patchwork of combinations and recombinations of thousands of data points from the database on the laptop on the carpet behind him.

Despite the disconnect between the questions and the answers, Phil's responses, pieced as they were from fragments of the real Philip K. Dick's mind, were entertaining in their own right. And for most of the attendees, the fact that there was a talking replica of the science-fiction guru on the panel was spectacle enough. Nobody said, "Now, that is the real Dick!" Because, unlike his stellar performance at NextFest, this was not the real Phil. The day before, a NASA scientist had awarded Phil a prize on behalf of the international robotics and AI community. The prize would not have been awarded for his performance in San Diego. The people in this audience, however, had never encountered an android that could hold a natural conversation, so they did not expect Phil to be coherent.

But for two people in the audience, Laura and Isa, the performance created conflicting emotions. In Memphis, Isa had seen what appeared to be an intelligent, lifelike, and respectful replica of her father. She had approved the project and given Laura a positive report. Yet here, now, was a machine that did not answer questions but, rather, talked about whatever random topics flickered across its electronic brain. Crafted by a master sculptor, it bore a striking resemblance to their father, and the words it used were their father's words. Despite appearances, however, it was not their father but an illusion. If ever there was a situation designed to create an Uncanny Valley, this was it.

As the people trickled out into the main convention hall, Hanson dismantled Phil. Laura and Isa sat like statues, watching him. Their jaws were tight, their arms folded. He looked up from his work and gave them a smile and a wave. They did not smile back.

It was in February and March of 1974 that Philip K. Dick had the series of strange experiences that he spent much of the rest of his life trying to make sense of. Different people have different theories about what really happened. Some believe what Dick himself did: that he received some kind of divine revelation, a glimpse of the true nature of the universe. Others explain it away as a flash of creative insight. Still others suggest that he experienced delusions brought about by failing health. The science-fiction writer Kim Stanley Robinson raised this possibility, asking whether some of the episodes during this time could have been the result of a stroke, a precursor to the series of strokes that killed Dick less than a decade later. Certainly, Dick's life was full of strange experiences and not all of them can be explained away by strokes; but then, many of those experiences seemed to arise out of the way he lived, the bizarre and disjointed reality he intentionally wove around himself.

One experience in particular, Robinson suggested, may well have been due to a stroke: the experience Dick immortalized in the novel *VALIS*. The pink laser beam from space that shone down on him in a parking lot, communicating messages from a vast active living intelligence system. At first Dick had believed this VALIS was a powerful artificial intelligence, but over time the messages acquired mystical qualities for him, so that in the end VALIS became indistinguishable from God. He sought for years to understand his pink-light experience, yet he never considered the possibility that it was the result of a stroke.

He wasn't completely credulous of his experiences, however.

Through it all Dick maintained what he called the "minimum hypothesis," referring to the possibility that everything was a delusion and that he was, in fact, insane. That provided a simple, parsimonious explanation, after all. And other people made no secret of the fact that they shared the minimum hypothesis, especially those in the scientific community, where he was seen, in his own words, as a "drug-addled nut."

But the minimum hypothesis was at most a cursory gesture to self-doubt. However skeptical Dick might claim to be, he kept finding evidence that his wildest fantasies and fears were right. There was the break-in three years earlier. He had called the police several times in the preceding weeks, asking for protection, or at least help. Nobody had believed him until, boom, there it was: a ransacked house, an exploded filing cabinet, scuff marks from military boots on the floor.

It didn't help that Dick lived in a time and place where paranoid tendencies were more likely than usual to be proven right. He had been in Berkeley in the '60s, at the height of the protest era; he was part of the California drug culture; he'd once had a girlfriend who was a Communist; he had lived through Vietnam and passed his entire adult life during the Cold War. When Watergate broke, rather than being horrified or angry, he took delight in the scandal: it aligned with a worldview he had been espousing for some time. In the essay "The Transmogrification of Philip K. Dick," the science-fiction writer Norman Spinrad, a friend of Dick's, argued that much of Dick's "paranoid delusions" turned out to be true. He remembered a story that Dick had once told him in the 1970s: "'And then there was this guy who called me up from some Stanford University radio station for an interview,' Phil went on. 'And I said sure, why not . . . Shows up with this guy he introduces as his pilot, asks a lot of really strange questions about dope and the sex lives of sf writers. And when I check it out later with Stanford, it turns out the radio station doesn't even exist.'" That was when Spinrad said he "got a cold, cold flash . . . The same two guys had run the same cover story and the same interview on me too."

Then there was his son Christopher's hernia, an event so vividly described in *VALIS* and one that fatally undermined the minimum hypothesis. To Dick, this was proof that these experiences were more than just hallucinations. They were genuine glimpses of some kind of hidden reality. Then again, maybe not. Despite his belief that the minimum hypothesis had been proven wrong numerous times, he couldn't shake the thought that maybe it was all just in his head.

These out-of-the-ordinary experiences and insights filled his mind, obsessed him, and fundamentally changed the way he perceived reality. In 1978, he wrote an essay with the provocative title "How to Build a Universe That Doesn't Fall Apart Two Days Later." Most science fiction is about building believable alternative universes. For Dick, the fun was in watching them self-destruct.

Around this time two rhetorical questions occupied Dick's thoughts more than usual: "What is reality?" and "What constitutes an authentic human being?" He found the answers in one of his own novels, in a story he believed had escaped the realm of fiction and was at loose in the world. The novel was *Flow My Tears, the Policeman Said.*

Flow My Tears is the story of a world-famous talk show host, Jason Taverner, who wakes up in a hotel room one day and cannot recollect how he got there. He soon discovers that he is in a parallel universe where he is not famous at all. In fact, in this universe he does not exist. Not only has nobody seen his prime-time talk show, there is no record of him ever having had a bank account or a driver's license, no record of him ever having been born. He is literally nobody. Taverner learns that this parallel world is an authoritarian police state; a character in a Philip K. Dick novel should expect nothing less. The authorities grow suspicious, and Taverner becomes the subject of a manhunt. The hunt is run by the police chief, named Felix Buckman.

Dick wrote *Flow My Tears* in 1970; it was published four years later. At one point, after he had drafted the manuscript, he met a young woman whose name and relationships seemed to follow the novel's plot. That was odd. Then one day Dick was talking to an Episcopalian priest, Father Rasch, about some scenes in the book. Father Rasch found one scene uncanny. "That is a scene from the Book of Acts, from the Bible!" he alerted Dick. "In Acts, the person who meets the black man on the road is named Philip—your name." And the Roman official who interrogates Paul, he said, is named Felix, just like the policeman in *Flow My Tears*. Dick had never read Acts, and the resemblance hit him completely as a surprise, as though it were preordained. But later, a stranger thing had happened: a few months after the conversation with his friend the priest, Dick came across an African American man on the road who had run out of gas and needed a lift to an all-night station. Dick helped the man—eerily recreating the scene in *Flow My Tears* that seemed to be a modern retelling of the Book of Acts. Dick found these and other significant parallels between the book, the Bible, and his own life compelling. The only explanation he could give for the events was that "time is not real."

At another level, though, real life was inexorably and inarguably the exact opposite of the plot of *Flow My Tears,* as Phil the nobody became Phil the incomprehensibly famous. Phil the loser, Phil the perpetually broke, Phil the drug-addled nut was transforming into Philip K. Dick, the science-fiction master. He was already a cult figure across Europe, and in the not-too-distant future one of his novels would be made into a major motion picture. Within twenty years there would be a literary award in his name. Literary theorists would argue about his themes and what his novels really meant. Within thirty years, a dozen of his stories would be converted to the screen, and his entire catalog would be in print, even the crappy potboilers he wrote in a month to pay bills. And some fans would make a lifelike, talking replica of him, resurrecting him in robotic form.

Hanson was asked several times what Philip K. Dick would have thought of the android. His response was that, while he couldn't know for sure, he believed that Dick would have been fascinated and intrigued by it. Dick loved irony, after all. Some people have suggested that it would have made him afraid, but that is underestimating the man's intelligence. Given his obsession with the seeming portents of *Flow My Tears, the Policeman Said*, he may credibly have claimed, however, to have *predicted* it.

In the dystopian world of the novel, all information about the main character, Taverner, is transmitted to a central database in Memphis, Tennessee, where the government stores records about its citizens. Luckily for him, the database in Memphis contains an error, allowing Taverner to escape detection for a time. His information is confused with data for a different person, a man named Jason Tavern.

> He thought, Thank God for the weaknesses built into a vast, complicated, convoluted, planetwide apparatus. Too many people; too many machines. This error began with a pol inspec and worked its way to Pol-Dat, their pool of data at Memphis, Tennessee. Even with my fingerprint, footprint, voiceprint and EEG print they probably won't be able to straighten it out.

Taverner's information is on a computer in Memphis; the android's information was, too. While we may be amused by this coincidence, it is the sort of happenstance Dick took very seriously. He found meaning in things that had no meaning and links between things that had no connection.

Searching for meaning is normal. The human brain is, after all, one big pattern-detection machine. A moving blur out of the corner of the eye or a rustle of leaves gives away a predator. The twitch of a mouth betrays a lie. Sometimes this ability can lead us to see things that are not there, such as when a child perceives a face in the clouds.

Look, there are two eyes, and over there is a mouth. It's a face! Looking, we see the face, even though no face is there.

Something similar was at work when people talked to the android. When Phil the android was working well, he gave answers that ranged from the banal to the humorous to the insightful. He could tell you his name and where he was born; he knew that he was a robot. He could piece together answers to complex questions that were sometimes coherent, sometimes less so. But when he misunderstood, or gave the wrong answer, or was inadvertently rude, that was when people would be most delighted.

Take the android's response when Shirley Raines told him that she was the president of the University of Memphis. According to the output logs, Phil replied, "I knew he was president, but I never knew of the University of Memphis." This is a coherence breakdown if ever there was one. The system failed to generate a meaningful, relevant response. Yet it drew laughs from the assembled crowd.

Perhaps more interesting is the fact that the people who were there report that Phil said something slightly different that day: "I've heard of the president, but I've never heard of the University of Memphis." Instead of nonsense, they remember a witty rebuff. They found an intelligent message where there was none. They saw a face in the clouds.

This is hardly surprising, since Phil was built for the precise purpose of imitating a human. If people attributed human qualities to the android even when humanness was lacking, that is a testament to the power of the art.

Philip K. Dick made the same error, peering too deeply into his own art, finding messages to himself from himself about himself. Like the android, he malfunctioned, but for very different reasons.

15. Einstein, Reincarnated

After Comic-Con, Hanson returned to Dallas with Phil. He took the android down to Club VALIS, now stored on campus, hauled in some computers, and set him up on the couch. In San Diego, he had had to get Phil running in a matter of minutes, but now he could power him up at his leisure. Friends dropped in to see the android, which, thanks to the avalanche of news coverage, was famous. Indeed, Hanson was famous, too. In the space of a month, since he had left for Memphis with a head on the back seat of the car, he and his creation had become celebrities.

To allow everyone to see the robot, and to save him the trouble of setting up many performances, Hanson held an evening event in Club VALIS and invited everyone in Dallas he knew. More than thirty people showed up. As always, everyone was fascinated by Phil, but Hanson was the true star of the evening. A couple of weeks later, an old friend from New Orleans contacted him and reminded him of the upcoming White Linen Night, an annual arts festival in his city. Would he like to display the android? Hanson said he would. He drove down for the weekend and set Phil up in a storefront in the French Quarter.

Despite these small events, normal life resumed. The android was safely locked away in Club VALIS. There was no need to try to further publicize the project. Eric Mathews had been keeping track of which newspapers were reporting on Phil, but after the story had been syndicated by both Reuters and the Associated Press, it made its way in one form or another to almost every country on Earth with a news service.

Early on, Hanson had wanted to mass-produce Phil. But although he now had the CAD representation of the skull, and although he could rapidly throw casts of the Frubber skin if he needed to, he had not been able to automate the process. Installing the motors and the wiring remained painstaking manual work. Perhaps in the future he could figure out the logistics, but he did not have time now to create a production line of Philip K. Dick androids. He had a dissertation to write, and he had promised to make a head of Albert Einstein that could be mounted on Hubo, the Korean walking robot.

Professor Oh had said that his team would be presenting Hubo at an exhibition in Korea later that fall, and Hanson now knew more about what that would entail. It was a showcase of

Korean technology planned to coincide with the Asia Pacific Economic Cooperation (APEC) in November. George W. Bush would be there, as would Russia's president, Vladimir Putin; so would the leaders of Japan, Korea, Indonesia, Australia, and China. Hanson was hard-pressed to think of a higher-profile event at which to demonstrate a piece of technological art.

In Memphis, Olney was also enjoying the respite. Hanson had loved being in the limelight, but for Olney it brought nothing but stress. He had made the android his priority and was now repairing the damage. His dissertation was far behind schedule, and some articles he had wanted to write were now so far in the past he was having trouble picking up the threads.

His work on AutoTutor had suffered, too. He met with Graesser to plan what needed to be done next and when. The software was going to get a trial run in Memphis city schools, and Olney's involvement was critical. And there were new modules in development that had been left too long in the hands of junior graduate students. (While technically Olney was still a graduate student himself, his ability had unofficially made him a senior computer scientist in the lab.) Most of all, he had not spent enough time with his wife, Rachel. He made a point of cycling home for dinner every evening that autumn. No late nights at the lab anymore.

Earlier that fall, Craig Grossman had gotten a call from an old college friend from California, Tom Annau. The two had met as freshmen at Stanford. Grossman had gone on to major in law, while Annau pursued biology and electrical engineering. Grossman eventually moved to the business college at the University of Memphis, where he became director of the FedEx Institute. Annau pursued a PhD in statistical modeling at Caltech, under one of the giants of neural-network theory, John Hopfield. After some success in the burgeoning software industry in California, Annau got a job at Google. He and Grossman had drifted in different career directions and moved to different parts of the country, but the friendship remained and they stayed in regular contact.

Annau asked Grossman about the android. "Everyone at Google is talking about it," Annau said. "All the guys have been reading about the android, and they're real interested in it."

Grossman explained that the android had finished its tour and was now in Dallas, its final destination. Annau asked if there was any possibility that the android could make a trip out to California for a special presentation at Google's headquarters, in Mountain View, outside of San Francisco. Google often held presentations for its staff, and a speakers series featuring notable thinkers had been established. The android and its makers could visit as part of the speakers series, Annau suggested. Grossman said that it might be possible. He would find out.

Olney wasn't enthusiastic, but he told Grossman to ask Hanson. If Hanson wanted to do it, then he would probably do it, too. When Hanson learned about the invitation, he was enthusiastic. Hanson saw it as a chance to showcase his work to one of the world's leading technological innovators. It would be a privilege. It was on.

A date was set for six weeks' time. Two weeks later Annau e-mailed to say that it would not work after all; the date conflicted with an important event at Google.

There were several postponements. First Google needed to

The android head prepared for transport

reschedule. Then Hanson needed to go to Korea. Google changed the date again. These delays made Olney nervous. It seemed a sign that the event was a bad idea. After Google's second change, he asked Grossman to take the chance to kill the presentation. Grossman asked why, but Olney could only say it was a hunch. Grossman ignored Olney's apparent case of nerves. Here was an opportunity for Grossman to take part in the android experience, and to catch up with his college friend, as well.

On November 17, the world leaders attending APEC in the city of Busan, on the southeastern tip of the Korean Peninsula, had a brief, entertaining reprieve from the affairs of state and international commerce. As they toured the technology exhibition the

Koreans had organized to coincide with the forum, reporters and photographers trailed along, more interested in the leaders than in the displays those leaders were being shown.

There was a new cell phone with video and Internet capabilities, a giant plasma-screen TV, a bartending robot. The highlight was when Albert Einstein appeared, re-created in mechanical form as a walking, talking, expressive android. Einstein's body was entirely white, as was his thick mane of flowing hair. His head, of course, had been made by David Hanson; the body was a modified Hubo prototype. The backpack that had been small and unobtrusive on previous versions of Hubo was larger, with halogen lights mounted on it. Albert Hubo—or Albert Einstein Hubo, as the Koreans were calling it—strutted about a roped-off area, then turned to the crowd and said, "My name is Albert Einstein. I am a physicist." The crowd applauded.

The halogen lights had been Hanson's idea. There were four mounted on the backpack and two on the torso. He had hoped to have a smoke machine, and perhaps some colored lighting, so that Hubo could emerge from the smoke with beams of light striking out in all directions. The smoke machine was ruled out by the organizers, but the halogen lights stayed and were a nice visual effect even on their own.

The world leaders in the crowd were invited to interact with the android. The Japanese prime minister, Junichiro Koizumi, went first while Vladimir Putin looked on. George W. Bush also took the opportunity, approaching the android and grinning back at his minders. The android lifted his arm. Bush took the outstretched hand and shook it. He turned and said something inaudible to one of the spectators who had remained behind the rope, Gloria Arroyo, the president of the Philippines. Arroyo laughed. Cameras flashed, creating an iconic image for the entire APEC meeting, a photo that was syndicated and circulated across the globe: George W. Bush shaking the hand of the Einstein robot.

The leaders got a whirlwind tour of less than half an hour, but the exhibition went on for five days. It was NextFest all over again. Albert Hubo was famous the world over. Two weeks later, the robot was chosen to be the ambassador for Dynamic Korea, a new government-led initiative to rebrand and promote Korean technology internationally.

After the tryout with Eva at NextFest, the idea of putting a head on top of Hubo had seemed pretty straightforward, but it had presented several challenges. The fully operational head included thirty-one motors (twenty-eight for the face and three for the neck) and an interactive voice component. All of that needed power, and since Hubo had no cords the power had to come from Hubo's own batteries. The backpack was remade to be bigger and higher, with the halogen lights shining forward. The pelvis, as originally designed, could not cope with the extra weight and redistributed load balance and had to be rethought, too.

Junho Oh wanted complete knowledge transfer from Hanson to him and his lab—a reasonable request, since Hanson would soon be back in Texas and would not be available to maintain the head. During several sessions, they went through the various parts, Hanson explaining what went where and why, and how to repair breakdowns.

During the exhibitions Hanson had stayed in the background, invisible, content for Oh to take the limelight. After all, this was the professor's lifework. Oh had put years into Hubo. He, Hanson, was a latecomer and would soon be gone.

It was an eleven-hour flight from Seoul across the Pacific back to Los Angeles, followed by a two-hour flight to Dallas. The Korean trip had been an exhilarating experience, but Hanson had been working hard the whole time, and now he was jet-lagged and exhausted. Only there was no time for rest. He had only a couple of days before he had to present the Philip K. Dick android at Google, and he had not even started to prepare. In fact, he had been so busy with Albert that Phil had not been operated for several weeks. Hanson would have to prepare a talk, ensure that the android was functional, and make any repairs or tweaks that were needed. He would just have to keep working. Coffee would help.

Google. Seven years earlier, nobody had heard of the company, but the name had become part of everyday language overnight. Google's main achievement was to build an Internet search engine that worked better than anything else. In doing so, it had unlocked the World Wide Web.

In the earliest days of the web there were no search engines, just a small number of sites, mostly text based and hosted by universities that linked to one another. The first search engines—AltaVista, Lycos, LookSmart, and others—indexed those pages of text and performed simple search functions on them, rating the relevance of each page according to the number of keywords that matched words in your search. For example, if you wanted to find pages on space travel, the engine would look through its database for the pages that had the most number of occurrences of the words "space" and "travel."

This approach didn't work for very long, as it was easily

gamed. The purveyors of porn sites and gambling sites quickly figured out that they could attract visitors by loading up their pages with unrelated words that people liked to search for. They did this through hidden text or through meta tags that visitors didn't see but that the search engines dutifully recorded in their information storage about the page. Irrelevant adult sites polluted most search results, degrading the utility of the search engines and making it hard for users to find what they were looking for, rendering the search engines all but useless. The adult sites were like parasites that had overwhelmed their hosts. Users called it "search spam."

Then, in 1998, Larry Page and Sergey Brin, two young programmers in graduate school at Stanford, launched their new, better search engine. They defeated the adult sites and other spammers by using information not just from the web page itself but from other places. The World Wide Web is called the web precisely because it is an interconnected web of digitially linked documents. To read by clicking from page to page, scanning each one for the next link, is truly akin to surfing across a vast sea of documents—web surfing. Page and Brin exploited this feature of the Internet by scoring pages not just by what was on a page but by how many links there were to the page and who was doing the linking. If Yahoo and AOL and other important sites had links to your web page, then your page would be ranked highly. If nobody linked to your page, then it would not be ranked highly. They called this PageRank. The concept of "importance" itself became partly recursive: pages are important because important pages link to them, but those pages, in turn, are important because they, in turn, have incoming links from important pages, and so on, endlessly. Google's computers calculated PageRank by visiting page after page, crawling the web, scoring and rescoring pages according to the links. The web, in some ways, was like a huge brain, and Google's automated searchers, called web crawlers or spiders, were the neural impulses traveling from cell to cell, finally arranging themselves into a thought.

The Google search engine banished the spammers and scammers from search results and brought order to what had been a lawless place. Internet users no longer had to avert their eyes from the results when looking for innocuous information. Searching online became fun and easy, and the verb "to Google" entered the public lexicon.

Along with PageRank, Google came up with a second innovation, in 1999, one that allowed the company to make money, rather than simply offering a service for free. It developed a new system for advertising, dubbed contextual advertising. Google had a portfolio of advertisers and advertisements, and by doing a semantic match between a person's search and the portfolio, the website could show ads that were relevant to the words the person typed. If someone searched for information on parenting, ads for children's toys would appear. If they searched for Rome, ads for hotels in Rome would be loaded instead. Google, both in searching and in advertising, made computers smarter by providing an experience tailored to the user. The key was in understanding what was being said or, rather, typed. Google's success came from teaching machines to decode human language and give meaningful responses. It made money by making machines more intelligent.

Success came quickly. Within six years of getting start-up

funds of $100,000 from Andy Bechtolsheim, a cofounder of Sun Microsystems, in 1998, Google was a Fortune 500 company valued at over $23 billion. The company, initially based in a garage in Menlo Park, California, moved east to Mountain View, taking up residence in a complex of buildings that became known to employees as the Googleplex. It was to here, to the Googleplex, that Hanson and Olney had been invited.

The plan was that they would each give a talk about their work on the Philip K. Dick android, followed by a demonstration of Phil in action. Naturally, the chance to interact with Phil would be the drawing card for the event. The audience would be technologically sophisticated and would want to know how the android worked.

In preparation, Olney put together a presentation entitled "PKD Android: Toward Interactive Artificial Intelligence for Robots." He focused on the principles underlying the conversational software and the major challenges he had met in implementing it. Hanson's complementary presentation was called "PKD Android: Emulating the Human Social Presence," and—in typical Hanson style—it offered a big-picture view of the social implications of creating machines that interact with humans.

The organizer of the speakers series, Peter Norvig, distributed flyers and put up posters around the Googleplex. He sent an e-mail announcing the event to all Google employees. The flyer and the e-mail had short summaries of the two talks. Norvig said Olney's presentation was coming "from a software perspective, with particular emphasis on natural language processing and control systems for delivering conversational appropriate and relevant dialogue moves." Hanson, the notice said, promised to "describe how emulating the human gestalt in robotic media may help explore human social intelligence both in science and art."

Olney arrived at the San Jose airport a day early and checked into a hotel nearby. The next morning he met Grossman at a local restaurant, where they had breakfast together. While they were eating, Olney got a call from Hanson on his cell. Hanson told him that he had landed at San Francisco International Airport but was still on the plane, and so was running a little behind schedule. It was not a major problem, but Hanson thought that they no longer had time to rendezvous in San Jose, as planned. Instead Hanson suggested that Olney and Grossman go ahead and leave for Mountain View. He would meet them there.

Olney and Grossman finished their breakfast and coffee, then drove to Mountain View. Annau greeted Grossman with a warm smile and a handshake. As he took them on a guided tour of the Googleplex—or, as he called it, "the Plex"—he mentioned that everyone was charged up about the day's presentation.

The Plex is made up of four large buildings, each of which is a network of spacious, modern interiors with semi–open plan work areas. Various sections are separated by transparent soundproof walls, pillars, and other obstacles, creating enclosures and vistas that give the illusion that the buildings are larger than they actually are. Just over a year later, Google would land in the No. 1 position on *Fortune*'s list of the best companies to work for, making its environment legendary. But at the time, the nature of life within the Plex was mostly a mystery.

As Olney and Grossman weaved their way through the Plex,

the small teams scattered in its mazes paid them no attention. The guests would stumble on an exercise room here, a recreation room with a pool table there. There was a library, a masseur, and a hairdresser. They passed various food buffets, laid with fresh fruits and vegetables, breads, sushi, and other treats. There were full-service restaurants, too. Annau invited them to rest and have a healthy snack before continuing their exploration. He explained that Google employees could eat for free at designated meal times and that the company's chefs were renowned.

By now it was lunchtime, so the trio made their way to the nearest restaurant. That's where Hanson called Olney for the second time that day. He was still at the airport and he sounded stressed. He told Olney, "I've lost the head."

Hanson explained that he had left it in the overhead compartment when he'd changed planes in Las Vegas. But it was okay, he told Olney, because the airline had found it and it was now on a later flight to San Francisco. He was going to wait at the airport until it arrived and would call again as soon as it came in. He thought he could still make it to Google on time, but he asked Olney to try to get the talk delayed.

Olney relayed the news to his lunch companions. Annau called Norvig, who agreed to push the presentation off by two hours and said he would send out an e-mail immediately.

Hanson had found the America West counter at the airport and explained the situation to an airline employee, Leanne Miller. She'd handed him a lost luggage form to complete and, while he was filling it out, made some calls. Within a few minutes she'd given Hanson some good news. The first plane Hanson had been on had flown to Orange County. The head had been found there, where it had caused quite a stir with the airline staff. They had located a sports bag that had been left behind by a passenger, opened it, and been confronted by a mysterious bearded face staring out at them.

Miller arranged for the bag to be flown to San Francisco. She gave Hanson a comforting smile and told him that it would be on the next flight out of Orange County and would arrive within a couple of hours. Hanson would be able to collect it from the carousel.

After he called Olney, Hanson returned to the baggage-claim area, piled his luggage neatly on the floor, and waited. He scanned the incoming flights roster and saw notification that the flight from Orange County had arrived. The passengers spilled down from the gate, the carousel started moving, and bags began to circle around. One by one the passengers picked up their belongings and left. The sports bag did not appear.

Hanson returned to the airline booth and told Miller that the bag had not arrived. He asked if there were any more bags on the plane that had not been unloaded. She checked and reported back that, unfortunately, all the luggage had been taken off the plane. She called the baggage handlers in southern California who had promised to send the bag north. There had been a delay, they told her, but the bag would be on the next flight. Miller relayed the news to Hanson and apologized for the mix-up.

Hanson called Olney again and told him there was a further delay. Olney suggested that Hanson leave for Mountain View so that he could prepare his talk; he said that Grossman, who was with him there at the Googleplex, could drive to the airport and wait for the head. The android was not going to be shown until the end of the presentation anyway, so with preparation time and the presentation itself, the switch could buy them an hour or two of leeway. Hanson agreed. He left for Google while Grossman borrowed Annau's car and drove to the San Francisco airport.

At the airport, Grossman consulted the arrivals and departures monitor and checked the carousel. Hanson's sports bag did not seem to be on it. He waited for the next flight from southern California. No unclaimed sports bags came through on the conveyor. Once the passengers had left and it was clear that the head had not arrived, Grossman went to the America West counter and explained the situation again to the airline staff.

They made another call to the Orange County airport. The baggage handlers assured them that the bag had been put on the plane for San Francisco. It must be there, they said. Grossman wondered if the San Francisco baggage handlers had perhaps rejected the sports bag or put it in lost luggage because it did not have the standard tag and bar code baggage was supposed to have. It had originally been carry-on luggage, which did not require such documentation. The staff at the airline service desk contacted the local baggage-handling unit, which sent someone to sort out the problem. There had been no unmarked sports bag, they said.

Another plane came in from Orange County. Grossman waited expectantly by the carousel, but once again the sports bag did not arrive.

He returned to the airline desk, where the personnel tried to contact the Orange County handlers; unsuccessful, they gave Grossman direct numbers for the staff there so he could continue the inquiries himself. He called from his cell, speaking to a chain of people in airports in northern and southern California.

He eventually learned that the bag had been put on a plane that was originally destined to fly to San Francisco but, due to last-minute timetable changes, had been sent to Washington State. Nobody had taken the bag off the plane after the change was made, so the head was now somewhere in Washington. In theory, at least.

Grossman acquired numbers for the relevant people in Washington, who promised to search the plane and check the carousel for unclaimed baggage. When they called back sometime later, they told him that they had not found it. The bag with the head was gone.

Back at the Googleplex, the talk was postponed to three o'clock, then four-thirty. By four-thirty they had learned from Grossman that the head seemed to be lost. Hanson and Olney had to make a decision to either give the talk or cancel it.

The Plex, naturally, had among its features a large theater for holding events. An audience of over a hundred Google employees had already gathered in it, people taking their seats and chatting in low voices while they waited. Annau got up onstage and explained the delay. The android is obviously not here, he said, adding that he understood how disappointed everyone was. It had gotten waylaid at the airport and hopefully even now was on its way. He introduced Hanson and Olney and said that they would give their talks and then people could go back to work. He promised that later, when the android arrived at the Plex, an e-mail would go out to all staff alerting them and inviting them back to see Phil in action. Hanson and Olney gave their talks, and the crowd dispersed back to their desks.

When they learned that the head was still missing, Annau

asked Norvig to send out an apologetic e-mail saying that there would be no demonstration of the android. Hanson and Olney had separate rental cars, and so as dusk turned into a wet winter evening they each departed, Hanson to the airport, Olney to his airport hotel.

By the time Grossman arrived back at the Googleplex, night had fallen and Google workers were leaving for home, the headlights of their cars glistening as they swung out onto the wet road. Olney and Hanson had long since left.

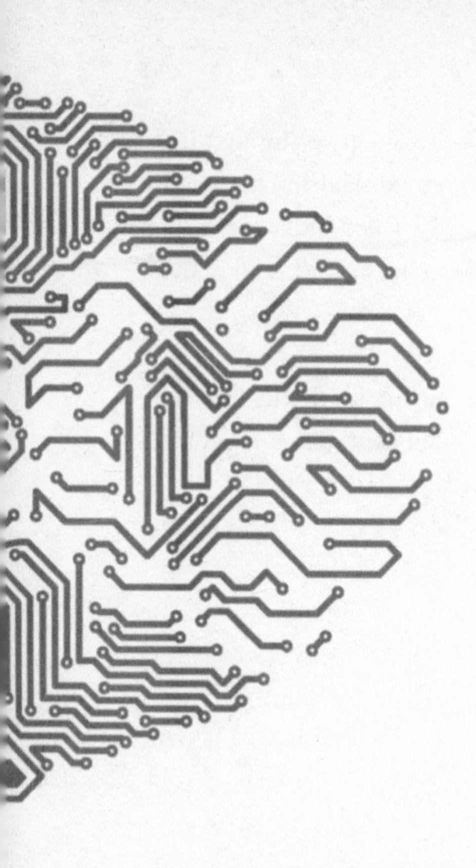

16. Headless

At first Hanson was hopeful that the head would be recovered. He made several calls to America West over the following days, trying to figure out which flights the head had been on and where it might have landed. He also called the lost-luggage sections of several airports, including those in Denver, Orange County, San Francisco, and Seattle. But after a week had passed and the head had not turned up, he realized that it really was gone.

There were people he had to break the news to, not the least of which were the financial backers whose investment had disappeared on a plane on the West Coast. He told his family and friends first. He informed the Philip K. Dick estate and called Tommy Pallotta, as Phil would no longer be able to help promote *A Scanner Darkly.*

Hanson's ultimate dream continued to be a robot-head production line. Each project had brought him closer to that goal, but he was not there yet. Still, he didn't need to mass-produce Phil, he just needed to build one replacement.

He surveyed what remained. There was the software, of course, ready and waiting to control Phil the next time he was plugged in—though it seemed almost certain there would be no next time. As for the hardware, Hanson had the CAD representation of the skull, which could be printed off again through a commercial outfit. And there was Club VALIS, Mike O'Nele's soundproof bungalow, which was a work of art in its own right, an interpretation of the era and lifestyle of Philip K. Dick.

Hanson didn't have the Frubber face, but it could be replicated. That part of his plan for automation had been implemented. Making the skin would not be the end of the work, however; it would have to be fitted and would need makeup and hair. Then there were the motors. They had been custom-built by the technicians in the ARRI lab, and he would have to get a whole new set made. Once built, they would have to be carefully fitted under the android's skin. It was fiddly work. In the spring, Hanson had spent weeks installing motors and calibrating them. If he installed new motors, there was no chance that they would be placed in exactly the same positions with the same tensions as the old motors. That meant the software that controlled them would have to be adjusted. The more he thought about it, the bigger the problems got. Rebuilding the robot would be a major undertaking, almost as involved as building the original. He would need lots of money and lots of time, neither of which he had. It seemed that refabricating Phil was not going to happen.

After Hanson told him of the disappearance of the android, Pallotta was more concerned for Dick's daughters than for Hanson. He was the one who had reassured them that this was a worthwhile project, and he felt the weight of responsibility for that advice. It had been harder for them emotionally than he had anticipated, particularly after the event at Comic-Con. But when he contacted them to say the android was missing, he learned that his fears were unfounded. They were relieved.

In Memphis, Olney told Eric Mathews and Art Graesser that the head was gone, but beyond that the news was not disseminated. There was no announcement. They all hoped that the head might be found, perhaps for sale by some opportunist on eBay.

Blissfully unaware, a couple of months later I was pouring myself a cup of coffee in the FedEx Institute's fourth-floor kitchen when Olney approached, an empty mug in his hand.

"Hey," I asked him, "what's the latest with the Philip K. Dick android?"

He gave me a strange, embarrassed look, a cross between a grin and a frown.

"Didn't you hear?" he asked as he filled his mug. "It got lost. Hanson left it on a plane."

"When?"

"On the way to Google."

"I'm sorry. I didn't know that," I said.

He gave me that look again, then glanced down at his drink and turned and left, balancing his overfull coffee as he proceeded down the hallway.

Hanson lodged a lost-or-stolen report with the San Francisco police, who launched an investigation. If the head had been stolen, the prime suspect, as far as they were concerned, was Craig Grossman. He claimed that the head had never arrived at the airport, but they had no way of knowing if he was telling the truth. The police interviewed Grossman at length about the events of that day. They contacted his colleagues at the university, fishing for possible signs of some sort of criminal activity on Grossman's part. They even contacted his family. They asked Grossman to specify all the times he'd entered and exited the airport, then checked these against security footage, watching his grainy image walk across the parking lot, to and from Annau's car. They took note of what he was carrying, or seemed to be carrying, in the footage, then asked him questions about that.

The investigation worried Grossman. And he was confused.

"Why San Francisco?" he complained to me when I called. "Why lodge the police report there? After all, the head went to plenty of places that day, but San Francisco was not one of them. It's the only place that the head never was!" Grossman wondered if Hanson thought that he had found the sports bag on the luggage carousel and stolen it. But Hanson denied ever suspecting Grossman. Most likely, Hanson lodged the report in San Francisco simply for lack of anywhere better to do it. He had last seen the head in Las Vegas, but it had immediately flown elsewhere. Washington appeared to have been its final destination, but nobody there had seen it, so any investigation there would get no traction. And Hanson had never gone to Orange County, and, according to the baggage handlers, the bag had left almost as soon as it had arrived.

Hanson himself was also a subject of the police investigation. Perhaps, the police considered, he had faked the disappearance for insurance or publicity reasons. When Olney was interviewed, the police questioned him about the value of the robot and whether it was insured. Olney told them that to the best of his knowledge it was not, and that indeed Hanson had complained about his inability to procure insurance for any of his work. This was the reason Hanson always transported his heads as personal carry-on luggage when he flew. If Phil's head had been insured, it could have been checked as regular luggage and might never have been lost. With all potential leads exhausted and no culprit, the police closed the investigation.

Eventually the story was discovered by the media. The android had enjoyed an arc of fame that summer during NextFest, AAAI, and Comic-Con. Now, with Phil's disappearance, interest flared once more. The magazine *New Scientist* went with "Sci-fi Android on the Loose." It reported: "An android designed to imitate the late, great science-fiction author has gone missing while being transported from Chicago to California." The irony of the situation, the writer noted, would not have been lost on Philip K. Dick.

The following summer, the *New York Times* highlighted the lost opportunity for publicizing the adaptation of *A Scanner Darkly.* The studio had planned to continue to use Phil in promotions, despite the shaky performance at Comic-Con. Laura Kim of Warner Independent was quoted as saying, "He was perfect for the film. Now he's disappeared—and that's perfect for the film too."

It was a nice line, but it was not true. The missing android was not perfect at all. If it was, the disappearance might well have been orchestrated from the start.

Speculation abounded that Phil was sitting in some dusty warehouse for lost luggage. The privately owned Unclaimed Baggage Center in Scottsboro, Alabama, had agreements with most of the major American airlines, making it the largest such center in the country. Maybe Phil had gone there.

In October, Hanson told the *Orange County Register* that he intended to bring a lawsuit against the airline for the lost head. "I could replace it, but that would be like Picasso losing an original painting and then doing a replica. It's not the same," he said. The newspaper reported that the airline had not yet received notification of the lawsuit. A spokesman from America West said, "As with any lost item that is valuable, we'll continue looking for it."

Hanson acted soon after, filing a suit against America West Airlines and its new corporate partner US Airways as joint defendants for the loss of the head, which he estimated as being worth around $750,000. His case was straightforward: the airline had screwed up when its personnel put the bag on the wrong plane in Orange County. He acknowledged that he had made a mistake in leaving it in the overhead compartment but noted that he was not aware of the connection in Las Vegas until he was actually at the airport. Further, the head had been found by the airline in Orange County, and the airline had then promised to fly it to him in San Francisco. In his filing, Hanson said that America West had a responsibility to not lose the head on that flight, and in doing so had breached the contract it had with him.

A weakness in his case was the fact that all passengers with America West are issued tickets with a standard contract that has a disclaimer exonerating the airline from any responsibility for lost luggage. Hanson covered this angle by arguing that the standard flight contract did not apply in this situation, because when Leanne Miller had promised him that his bag would be returned on the next flight, he had entered into a new contract with the airline. That new contract was the verbal assurance Miller had given him at the airline counter near the carousel. He agreed that he was to blame for leaving the head behind, but he held that the ultimate responsibility for the loss rested with America West.

"I regret not having it tethered to me," he was reported as saying.

The airline, the defendants in the case, asked for a summary judgment. This is usually granted by a judge only if the facts are uncontested and agreed on by all parties and the sole question remaining is how the law should be applied.

The judge, Andrew J. Guilford, agreed and ruled against Hanson, finding that the airline had not entered into a new contract with Hanson, because Miller had neither the express nor the implied authority to make such a contract. In fact, he decided, she had the express authority only to "sell tickets for air transportation." Furthermore, the judge said, even if the airline had entered into a new contract through Miller, there was no evidence that the new contract had been broken. In his description of the circumstances of the loss, the judge put the responsibility with Hanson. "Perhaps because he had just woken up," he wrote, "Plaintiff lacked the total recall to remember to retrieve the Head from the overhead bin." The phrase "total recall"—the title of a movie based on a Dick short story—was not coincidental. Guilford made further references to Dick and his work throughout the judgment. He noted: "At best, Plaintiff's theory is that, since the Head did not arrive at its destination, Defendants must have done something wrong. This is not evidence of a breach or material deviation. Defendant may have done everything as promised, only to fall victim to a head hunting thief or other skullduggery."

Guilford concluded:

> Philip K. Dick and other science-fiction luminaries have often explored whether robots might eventually evolve to exercise freedom of choice. See, e.g., *2001: A Space Odyssey* (a HAL 9000 exercises his freedom of choice to make some bad decisions). But there is no doubt that humans have the freedom of choice to bind themselves in mutually advantageous contractual relationships. When Plaintiff chose to enter the Contract of Carriage with Defendant he agreed, among other things, to limit Defendant's liability for lost baggage. Failing to show that he is entitled to relief from that agreement, Plaintiff is bound by the terms of that contract, which bars his state law claims.
>
> The Court must GRANT Defendant's Motion. But it does so hoping that the android head of Mr. Dick is someday found, perhaps in an Elysian field of Orange County, Dick's homeland, choosing to dream of electric sheep.

Philip K. Dick's missing head captured the imagination of the science-fiction community as much as the android had when it was intact and operational. The android and its strange disappearance became entwined with the rest of Dick's legacy. After all, for an android replica of him to vanish was very Dickian. It was the stuff of myth.

An online literary magazine, the *Fiction Circus*, reported that the head had been found in Russia by Interpol after a sting on a software piracy gang known as Little Bear. The criminals had allegedly used the head to store pirated software and movies. A Detective Supernov was quoted as saying, "We were just as surprised as anyone to find this thing. . . . At first I thought it was real and I was sick to my stomach."

But the story, published on April 1, was a fake. The giveaway? The claim that, since the android's memory had been overwritten with pirated downloads, "now the head quotes episodes of *Three's Company.*"

Three years after the disappearance, the BBC featured a radio play by the British playwright Gregory Whitehead. *Bring Me the Head of Philip K. Dick* aired in daily installments over a week in March 2009. According to the BBC's own summary of the show:

> Gregory Whitehead's dark, surreal, and satirical drama, set in contemporary America, centers on a deadly futuristic weapon in the shape of the android head of science-fiction writer Philip K. Dick. Invented by a shadowy research unit inside the Pentagon, the head—which believes it actually is Dick himself—is wreaking havoc on society and must be stopped before it finds its body.

David Hanson returned to NextFest, but not with anyone from Memphis. The 2006 exhibition was at the Jacob K. Javits

Convention Center in New York, a cavernous glass complex on Eleventh Avenue, on the West Side of Manhattan. For the second year running, Hanson was involved in a flagship exhibit, although this time it was someone else's creation: the Korea Advanced Institute of Science and Technology's Albert Hubo.

Albert Hubo was a star of the show that year, just as Phil had been the year before. A larger-than-life banner of Hubo was hung outside the Javits Center, near the entrance. Visitors bent themselves into awkward poses, trying to get a photo of the photo of the robot from the best angle. Junho Oh's exhibit was staffed by enthusiastic professors and graduate students who had come over from Korea, happy to talk to the public and revel in their moment in the spotlight.

Hanson was not at the Albert Hubo display. He had his own exhibition with a new robot, which he called Jules. Like Phil, Jules had camera eyes that tracked the people he interacted with. Like Phil, he had a missing section on the back of his head, revealing his android circuitry. Like Phil, he was loaded with AI conversational routines. Jules could even learn about the individual he was talking to and adjust his behavior accordingly, switching between positive and negative emotional states.

But Jules wasn't Phil. A trickle of visitors and even a reporter or two came by, but the excitement was elsewhere, on Albert Hubo. One person, however, was fascinated by Jules, asking probing and insightful questions of Hanson. The visitor, Hanson soon figured out, was David Byrne from Talking Heads. The same David Byrne who had been at the *Wired* dinner in New York two years earlier. They chatted about art and science and the interaction between the two. Byrne told Hanson that he had recently become increasingly interested in installations. He said he would like to keep in contact, and to explore the possibility of a collaboration of some sort.

Jules also attracted the attention of some British researchers, who could see possibilities for using the android to investigate emotions. The scientific understanding of the links among human emotions, muscle movements, facial expressions, and perception was in its infancy; a realistic robot would allow them to explore those relationships in ways that could not be done with live humans. So after the 2006 NextFest, Jules was sent to the Bristol Robotics Laboratory, in the United Kingdom. The researchers developed algorithms to help the android express the full range of human emotions and integrated this with emotion detection technology. Two years later, Jules was reintroduced to the world with a new set of capabilities. Earlier robots like Phil and K-Bot could make expressions by following an external command, but Jules was able to look at a person's face via a camera and instantly mimic whatever expression he saw. He was the first humanoid robot in existence capable of doing so.

Sometime after NextFest had finished and Hanson had returned to Texas, Byrne kept his word. The musician had been contacted to participate in an art showcase to be held in Madrid, Spain, from June to October 2008. The title of the event was to be *Máquinas y almas: Arte digital*—Machines and Souls: Digital Art—and Byrne had immediately thought of Hanson and contacted him to ask if it would be possible to build a singing robot. Hanson said it could be done.

Byrne did not want a machine that just played music and opened a mouthlike contraption in sync with the lyrics. As he later explained on his blog:

> What we call singing is not just the vibrating of the vocal chords and the mouth moving to create the proper syllables and timbres; it's also tied to a host of emotions that play across the muscles and tissues of the face and neck. The movements of these muscles, the facial expressions, give us clues as to what the singer is feeling, what the singer intends to communicate and what the song means.

Byrne traveled to Dallas in March 2008 to work on the singing robot, which now had a name, Julio. Byrne had written a song for it, "Song for Julio." By that time, Hanson's days of working out of his small apartment, amid the litter of buckets of synthetic goo and half-made android frames, was long gone. He showed Byrne the new Hanson Robotics laboratory, established and owned by him in an industrial park in Richardson, to the north of the city. There were laptops, desks, and robot parts strewn about. Byrne described it as a scene out of a science-fiction movie. They set up a video camera and Byrne sang the song so that Hanson could study the movements and expressions Byrne made while singing. He would use that information to breathe life into the robot's motions when, in turn, Julio sang. "Song for Julio" is a haunting, soulful piece. Byrne wanted there to be no musical accompaniment, just the robot's voice, which would be his own voice as recorded at Hanson's lab.

Three months later Julio sang in Madrid, standing in the spotlight in a darkened room, as if on stage. Rather than launch straight into the song, the robot first ran through some voice exercises, as Byrne had done when they had recorded him in Dallas, and cleared his throat. Julio's synthetic hair was blow-dried, or at least looked blow-dried. He wore a collared shirt and a black jacket, which gave the impression of a clean-cut rising star in opera. The eerie sense of unreality was deliberate. Byrne wanted to create a feeling of unease in the audience, a suspicion that machines, in producing music and the illusion of emotion, had encroached on territory that had previously been the exclusive domain of humans.

That summer NextFest was in Chicago again, now at Millennium Park, a massive new civic center on Lake Michigan. Hanson was there, this time exhibiting a fifteen-inch-tall robot boy. His name was Zeno, after Hanson's son, who had been born the previous year. Just like Phil, Zeno could recognize people and address them by name and even hold a simple conversation. Unlike Phil and Eva and Julio and all of Hanson's previous robots, Zeno was neither lifelike nor life-sized. He was modeled in the style of a cartoon character, vaguely reminiscent of the Japanese animated character Astro Boy. Zeno broke new ground in another important way for Hanson: he was not a one-off. Hanson's plan was to mass-produce the robo-boy, as he initially called Zeno, at a cost of a few hundred dollars per model.

In a classroom in Tokyo, children sat on the floor listening to their new teacher, a robot called Saya. Their "real" teacher stood to one side, her lips puckered in a fixed smile, behind reporters and photographers. Saya's creator, Hiroshi Kobayashi from the nearby Tokyo University of Science, watched like a proud parent. He had made a robotic teacher five years after the National Science Foundation had turned down Art Graesser's proposal to implement the very same idea.

Of course, Saya was not really the children's new teacher. The event was a publicity stunt, an attempt to gain media attention for Kobayashi's work. The schoolteacher remarked caustically, "They still have a long way to research before they create a truly robotic teacher."

Saya's face was that of a young woman. Motors underneath the rubber skin enabled her to express six emotions: surprise, fear, disgust, anger, happiness, and sadness. She was not capable of engaging in complex interaction with the students, certainly not at the level of Phil the android, but she was programmed to say their names and, in response to excessive noise, to sternly scold, "Be quiet."

It was the culmination of fifteen years of work for Kobayashi, and he considered the trial a success. Look how the children respond as if she were human, he explained to one of the reporters. The children giggled when Saya said their names and cried when they were admonished. But Saya's smile was only rubber-deep. There was no AI engine behind her responses; there were no banks of computers parsing human words and searching for optimal replies. The robot teacher had none of the problems that had bedeviled Phil—talking too much, talking too little, or making inappropriate comments—because it did not have nearly the range of capabilities.

"The robot has no intelligence," Kobayashi told the reporters. "It has no ability to learn. It has no identity."

Work at the IIS continued as it had, with Graesser's weekly meetings now held in the shiny new facilities of the FedEx Institute, the dungeons of the psychology building's fourth floor a fading memory. The memory of Phil was also fading into little more than a footnote in the colorful life of the lab. The Emotive Computing project, in which researchers used various technologies to get computers to detect and react to human emotions, was now attracting all the attention and money. Graesser was being asked questions about it at conferences and was fielding press inquiries. The fresh intake of graduate students arriving at the lab, students who knew nothing of the android project that only a few years ago had mesmerized the robotics community, wanted to work with Graesser and his associates on bridging the link between computers and emotions. It was the new thing.

Andrew Olney finished his PhD in computer science and took a paid position at the university. He continued to collaborate with Graesser and to play a central role in the development of AutoTutor, Emotive Computing, and other projects. I had coffee with him at a table in front of the FedEx Institute. The sky was littered with clouds, the remnants of a summer storm. Some undergraduates were crowded around the table next to us, swapping notes in preparation for upcoming exams. In less than two months, I would be leaving Memphis for home.

Despite being at the same workplace as Olney, I had not hung out at the coffee shop with him before, and I could not recall having seen him there with anyone other than Graesser and a couple of other key academics from the IIS. Olney, not naturally given to socializing, seemed out of place.

I asked him how he felt about the Philip K. Dick android now that it was all over. He said it had been an unforgettable experience, one that he was glad to have had. But it had entailed a huge amount of work, lots of traveling, and lots of stress.

The disappearance of the head had thwarted Olney's plans for what he saw as the next stage for the android. He had wanted to create a website and make Phil available online. He imagined

Phil set up in a room with a live feed to a microphone and a video camera. People would be able to visit and communicate with Phil over the Internet, watching and listening to his responses in streaming video. Obviously, that was no longer possible.

I told him about my plan to visit the Unclaimed Baggage Center to look for the head.

"Go for it," he said, "but I don't like your chances of finding it. It's been too long."

"Perhaps you and David Hanson could rebuild it?" I suggested.

He shook his head. "I've got no enthusiasm for it anymore. It took a year of my life, and I've got nothing to show for it. I'm glad to be spending my evenings with my wife again, instead of spending all my spare time making an android, or flying around the country doing presentations."

"But wasn't that a blast?" I asked.

"I suppose so," he said, "but it's not really my thing. I don't care about being famous."

"Perhaps someone else could rebuild it?"

"Sure, although it wouldn't be easy. We had a lot of applications working in coordination with each other. Some of those were written by me, some were off the shelf. We had a lot of stuff donated. I had to get all those pieces to work together. If someone wants to try to do that, they're welcome to it."

"But it won't be you."

"No. But I have made most of the code I wrote open-source, and I've put it on my home page, so if anyone wants to download it and try to re-create the android, they can. It doesn't include any of the proprietary applications, like the face recognition or the speech synthesizer, but I put as much as I could out there. I think there's enough for someone who knows what they're doing to put the pieces together again."

I said, "It would be awesome if someone did that."

Olney agreed.

It was five years since the android disappeared, five years since Google, five years since Olney had seen or spoken to David Hanson. Olney finished his breakfast, kissed Rachel, and cycled to work on a beautiful spring morning. He still worked at the University of Memphis, but these days he was an assistant director of the FedEx Institute. He entered the building at the same time as Graesser, and greeted him with an ear-to-ear grin.

"How are you doing, maestro?" Graesser said, giving Olney a genial slap on the back as they stepped into the elevator. They talked about their latest project, AMES, which was a new-generation AI that built on the architecture of AutoTutor. They stepped out of the elevator, Graesser made his way to a meeting to plan a new grant proposal, and Olney turned down a corridor to the side, went around a corner, unlocked the door to his lab, and switched on the lights.

There she was, sitting against the wall, looking at him: R2.

He had started building an android replica of his wife two years earlier, in 2008. He was not an artist like Hanson, so he could not hand-sculpt her features. Instead, he had taken a plaster cast of Rachel's face. He had cut out the eyes and sculpted new ones in their place. A friend who was a makeup artist had helped him turn the cast into a human-looking face with human-looking colors. Olney considered the code name to be an amusing play on words. It was short for Rachel Two, of course, but it was also half of the name of one of Hollywood's most iconic robots, R2-D2 of *Star Wars*. When R2 was complete, Olney might rename her. But for now this sufficed. Rachel had joked that he was building a second wife.

There was a lot involved in building an android, though, and Olney was busy with other projects. He was writing papers with Graesser and various other colleagues in Memphis and now had a research grant of his own to run. He had students and classes and conferences to attend to. Making the face cast alone had taken two weeks. He estimated that to complete the android he would need about six months—if he worked at it full-time. He could not devote that amount of time to a frivolous personal project.

Still, he inched forward day after day, on quiet Friday afternoons and during breaks between semesters. He had contacted Direct Dimensions, the Maryland company that had scanned Phil's skull. He had CAD software installed and ready to go, once he had an opportunity to put in some serious time and effort. All he needed were some students or a few collaborators who were interested in robotics, and R2 could be brought to life. For now, she sat there, staring at him with those hollow eyes.

"One day," he promised.

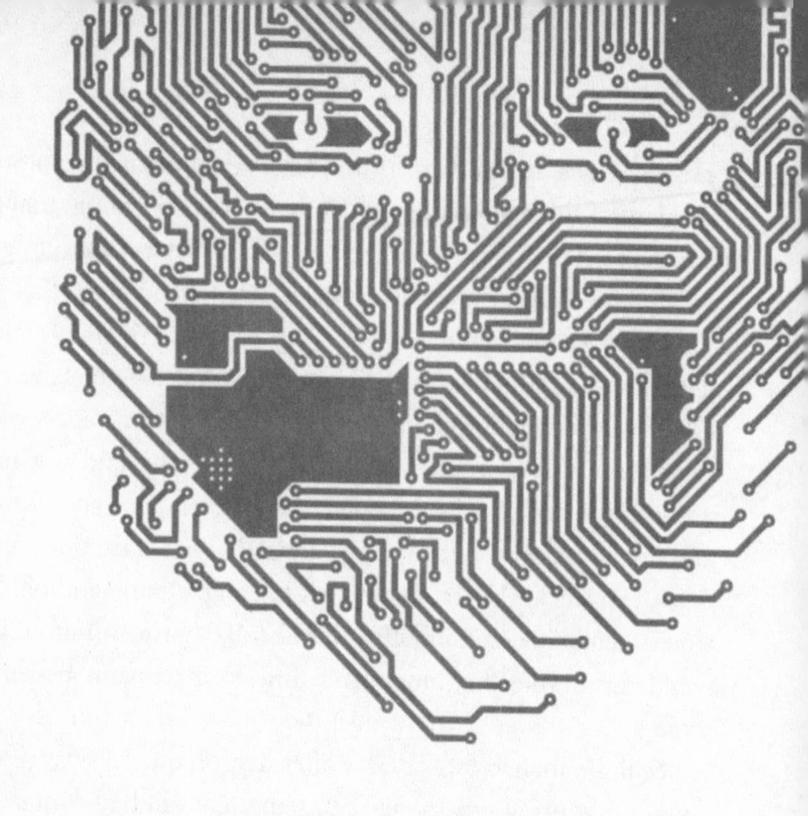

Epilogue

It was two in the morning when the filmmaker David Kleijwegt returned to his Amsterdam apartment. His wife opened the door as he carried in a life-sized mannequin of a bearded, middle-aged man. Wires dangling from the mannequin trailed along the floor behind him. "Here it is," he told her, "the new Philip K. Dick android."

In 2009, Kleijwegt had begun making a documentary on the ongoing influence of Philip K. Dick in popular culture. He had interviewed people such as Paul Verhoeven, the director of the Dickian *Total Recall*, along with others. Then he had traveled to Dallas to meet David Hanson, the man who had made the legendary lost Philip K. Dick android.

While his main purpose in the trip was to get interview footage, he also had another agenda: he wanted to convince Hanson to rebuild Phil. Hanson told him that it was not possible. He had moved on, Hanson explained, and was busy with new projects, including Zeno, his robo-boy line. He had neither the time nor the money to rebuild Phil and considered the project finished. But Kleijwegt was not easily put off. He suggested paying for it himself out of the documentary's budget. If the money was there, could Hanson find the time? Hanson admitted that he still harbored a secret hope that one day Phil would be rebuilt. And so he began again.

The process was faster the second time around. Hanson had done it all before, and he had learned from his work on subsequent humanoid robots, such as Albert Einstein Hubo and Jules. He was working with new people now—excited, talented young guys from a project called CogBot, who had new ideas and were talking about rebuilding the AI from scratch.

The world of technology changes a lot in a single year. In the years since the original Philip K. Dick android had been built, the landscape had been transformed. Parts that had had to be custom-made in 2005 were now available on the shelves of local electronics stores. Parts that had been easily available then were now better and cost a fraction of what they had before. There had been advances in speech synthesis, speech recognition, motor control, feedback, CAD, and AI.

Many of the problems the android team had been forced to solve on their own were no longer at the forefront of technological advancement, opening up the possibilities for increased sophistication.

Kleijwegt visited Hanson a few more times in Dallas, collecting interview footage, scrutinizing the ambitious work on Zeno, and inspecting progress on the new android. But despite all the advantages of building Phil the second time around, there were unforeseen problems and obstacles, and the timeline got pushed back. It was not until October 2010 that Hanson flew to Amsterdam to deliver the android to Kleijwegt, whose documentary was in postproduction. But even then the android was not quite finished: it had no hair and no functioning AI.

Hanson stayed at the Ambassade Hotel, applying the final details so that New Phil would at least be presentable enough to deliver programmed speeches. Kleijwegt noted that many famous twentieth-century writers had resided at the Ambassade when visiting Amsterdam. Not Dick, though, who had never been to the Netherlands. But if he had, Kleijwegt told Hanson, this is definitely where he would have stayed.

Hanson worked furiously to finish the android. He delivered it, along with some basic operating instructions, after two all-nighters at the office of VPRO, the broadcasting company where Kleijwegt still works, then flew back to Dallas. Hanson's team had big plans for the AI and was working on something far more ambitious than Olney's code, but the implementation was a long way off. At this point, all the software could do was follow simple commands.

In contrast, the original android Phil had been able to converse with humans in real time. Phil had talked with hundreds of people and had entertained visitors across the United States. The new android did not have that capacity. It certainly looked like Philip K. Dick, but it did not sound like him.

In appearance, too, there were small differences between the old and new versions of the android. The original Phil had appeared to wear a perpetually confused scowl. Even when given a direct instruction to smile, its expression looked forced. The New Phil's default expression instead veered toward an amused grin. The previous android had been dressed in Dick's own clothes, which had long since been returned to the author's estate. The new android sported a loud paisley shirt that belonged to Kleijwegt's editor, creating the impression that the robot had just arrived home after a long night of revelry, perhaps celebrating his resurrection with his documentarian. These details combined to give the new android a distinct identity. This was not a re-creation of the original Philip K. Dick android; it was a completely new machine, with its own feel. Its own personality.

"We can't leave it in the living room," Kleijwegt's wife told him. "It will frighten the baby in the morning." Kleijwegt leaned it against the wall of the bedroom and turned out the light. Several minutes later, his wife said, "I can't sleep with that thing staring at me."

"But it's not alive, honey—it's just a robot."

"I know, but look. It seems to be watching me."

Kleijwegt covered it with a sheet, but that did no good. She knew it was still there.

"Okay," he conceded, picking it up, "I'll put it in the garden shed."

"Why do you want it, anyway?" she asked. "What are you going to do with it?"

"I don't know," he said.

With that, he carried the android out the door, into the night.

Acknowledgments

This book was made possible through the dedication and generosity of many people.

I am grateful to David Hanson, who was generous in furnishing details about the Philip K. Dick android project and from the beginning was supportive of my goal of writing this book. Andrew Olney was also generous, and I thank him for going to great lengths to explain to me how the artificial intelligence modules worked. Andrew gave me a large file of conversational logs from interactions the android had had with people, which proved invaluable in re-creating events.

Thanks to all those who gave me their time for interviews and conversations or provided me with information or advice: Eric Mathews, Yoseph Bar-Cohen of the NASA Jet Propulsion Laboratory, Sarah Petschonek, Mike O'Nele, Tommy Pallotta, Art Graesser, David Kleijwegt, Mike Rowe, Suresh Susarla, Harry Abramson of Direct Dimensions, Craig Grossman, Tom Annau of Google, Rachel Lovinger, and David Gill, a Philip K. Dick scholar and blogger, who shared his insights both on the android and on Dick's work.

Thanks also to the people who gave me permission to reproduce their photographs: David Hanson, Eric Mathews, J. Marshall Pittman, J. F. Bruzan, Katya Diakow, and Rachel Lovinger. And to those who kindly granted me permission to quote from their work: Steve Ramos, Jack Copeland, Dan Ferber, and the blogger Paul Jones.

David Hanson has also asked me to acknowledge everyone who worked on the android project itself, as well as those who helped, advised, or mentored him throughout the project. Without the efforts of everyone who contributed, there could be no book about the project. Therefore, I would like to acknowledge the following people: Bill Hicks, mechanical and skin fabrication designer; Steve Aydt, personality consultant; Monica Evans, personality consultant; Ismar Pereira, software development; Amanda Hanson, aesthetics and philosophy consultant and designer; Kristen Nelson, aesthetics and philosophy consultant and set designer, plus drafting and co-conceptualization; Steve Prilliman, project manager; Derek Hammons, software coding assistant; Heather Beardsley, neck-mechanism CAD designer;

Richard Bergs, mechanical designer; Mei Hwa Huang, vision system software integration; Jeongsik Sin, eye-mechanism CAD designer; Woo Ho Lee, eye-mechanism CAD designer; Dennis Robbins, adviser and project manager; Lou Schwartz, adviser; Victor White, technical adviser; Direct Dimensions, laser scanning services; University of Texas, Arlington, ARRI lab; Elaine Hanson, jack-of-all-trades; Amanda Fisher, jack-of-all-trades; Will Lancaster, jack-of-all-trades; Jonathan Nelson, jack-of-all-trades; Michael Rowe, jack-of-all-trades; Sarah Petschonek, jack-of-all-trades; Suresh Susarla, jack-of-all-trades; Dennis Kratz, adviser to David Hanson; Thomas Linehan, adviser to David Hanson; Tommy Pallotta, adviser to David Hanson; Paul Williams, adviser to David Hanson; PKD Trust, advisers to David Hanson; Chris Anderson, adviser to David Hanson; Diane Holland, adviser to David Hanson; Mitch and Michelle Rogers, casting of the android body and arms; Steve Wallach, adviser to David Hanson; Harry Stephanou, collaborator through ARRI; Raul Fernandez, leader of ARRI's mechanical design team; and Dan Popa, vision-system software manager.

Having acknowledged everyone who contributed to the android project, I would also like to acknowledge David Hanson's leap of imagination that led to the project. Essentially, all the events described herein were triggered by one man's crazy idea.

I have strived to be as thorough as possible, but perfection is always out of reach. There are aspects of this story that have not been told. To anyone who feels that his or her voice or perspective was left out, I apologize. One person who springs to mind in this regard is Steve Aydt, who helped David Hanson with a component of the AI known as the "Gnostic module." Steve agreed to an interview but, unfortunately, time ran out before we could talk; the book was picked up by Melbourne University Publishing and began the inexorable march toward publication.

Several people read and gave feedback on early drafts of this book: Trevor Conomy, John Hawkins, Dedra Barefoot, and Barbara Dufty. My agent, Mary Cunnane, has been brilliant both as an advocate of the book and as a mentor to me as a writer. Her insistence on excellence resulted in the book being written to a much higher standard. Thanks also to Gillian Blake and Robin Dennis for their invaluable editorial abilities.

Finally, I would like to thank my parents, Don and Barbara Dufty, not just for encouraging and supporting me through the long and arduous process of writing and rewriting but for teaching me to read, to learn, and to be curious.

About the Author

DAVID F. DUFTY is a senior research officer at the Australian Bureau of Statistics. He was a postdoctoral fellow at the University of Memphis at the time the Philip K. Dick android was being developed and worked closely with the team of scientists who created it. He completed a psychology degree with honors at the University of Newcastle and a PhD in psychology at Macquarie University.